Multicarrier Techniques for 4G Mobile Communications

Multicarrier Techniques for 4G Mobile Communications

Shinsuke Hara
Ramjee Prasad

Artech House
Boston • London
www.artechhouse.com

Library of Congress Cataloging-in-Publication Data
Hara, Shinsuke.
    Multicarrier techniques for 4G mobile communications / Shinsuke Hara, Ramjee Prasad.
        p. cm. — (Artech House universal personal communications series)
    Includes bibliographical references and index.
    ISBN 1-58053-482-1 (alk. paper)
    1. Universal Mobile Telecommunications System.   2. Carrier waves.
    I. Prasad, Ramjee.   II. Title.   III. Series.
    TK5103.4883.H37   2003
    621.382—dc21   2003048095

British Library Cataloguing in Publication Data
Hara, Shinsuke
    Multicarrier techniques for 4G mobile communications. — (Artech House universal
    personal communications series)
    1. Mobile communication systems
    I. Title   II. Prasad, Ramjee
    621.3'8456

    ISBN 1-58053-482-1

Cover design by Igor Valdman

International Standard Book Number: 1-58053-482-1
Library of Congress Catalog Card Number: 2003048095

10 9 8 7 6 5 4 3 2 1

To my parents Kokichi and Kazuko, to my wife Yoshimi, and to our sons Tomoyuki and Tomoharu
—Shinsuke Hara

To my wife Jyoti, to our daughter Neeli, to our sons Anand and Rajeev, and to our granddaughters Sneha and Ruchika
—Ramjee Prasad

Contents

Preface

यदा संहरते चायं कूर्मोऽङ्गानीव सर्वशः ।
इन्द्रियाणीन्द्रियार्थेभ्यस्तस्य प्रज्ञा प्रतिष्ठिता ॥

yadā samharate cāyam
kūrmo 'ngānīva sarvaśah
indriyānīndriyārthebhyas
tasya prajñā pratisthitā

"One who is able to withdraw his senses from sense objects, as the tortoise draws its limbs within the shell, is firmly fixed in perfect consciousness."

—The Bhagvad Gita (2.58)

At recent major international conferences on wireless communications, there have been several sessions on beyond third generation (3G) or fourth generation (4G) mobile communications systems, where modulation/demodulation and multiplexing/multiple access schemes related to multicarrier techniques have drawn a lot of attention. We often met at the conference venues and realized that no book covered the basics of multicarrier techniques to recent applications aiming at the 4G systems. Therefore, we decided to write a book on multicarrier techniques for 4G mobile communications systems.

Figure P.1 illustrates the coverage of the book.

This book provides a comprehensive introduction to multicarrier techniques including orthogonal frequency division multiplexing (OFDM), putting much emphasis on the analytical aspects by introducing basic equations with derivations.

This book will help solve many problems encountered in research and development of multicarrier-based wireless systems. We have tried our best to make each chapter comprehensive. We cannot claim that this book is errorless. Therefore, we would really appreciate it if readers would provide us with any comments to improve the text and correct any errors.

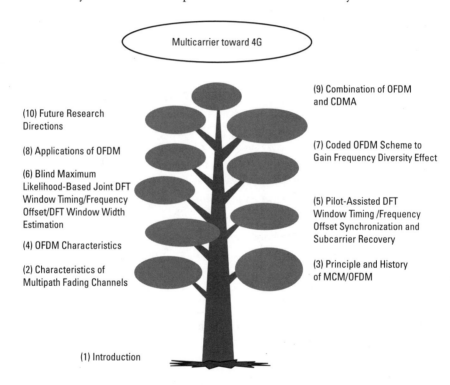

Figure P.1 Illustration of the coverage of the book. The number in branches denotes the chapter of the book.

Acknowledgments

The material in this book is based on research activities at Osaka University in Japan, the Department of Communication Technology at Aalborg University in Denmark, and Delft University of Technology in the Netherlands. The authors would like to thank Professor Norihiko Morinaga (Osaka University), who gave Shinsuke a chance to work with Ramjee in the Netherlands in 1995–1996. They also wish to thank Dr. Jean-Paul Linnartz (Philips National Laboratory), who also gave Shinsuke a chance to do research in the Netherlands. Their heartfelt gratitude also goes to Professors Seiichi Sampei and Shinichi Miyamoto (Osaka University), who kindly took care of Shinsuke's students during his absence.

They are deeply indebted to Professor Minoru Okada (Nara Institute of Science and Technology in Japan) and Dr. Yoshitaka Hara (Information Technology R&D Center at Mitsubishi Electric Corporation in Japan) for their discussion, interaction, and friendship with Shinsuke over the years.

The material in this book has benefited greatly from the inputs of the following many brilliant students who have worked with us on the topic: Kazuyasu Yamane, Kiyoshi Fukui, Ikuo Yamashita, Masutada Mouri, Tai Hin Lee, Frans Kleer, Daichi Imamura, Masaya Nakanomori, Shuichi Hane, Shigehiko Tsumura, and Montee Budsabathon.

Last but not the least, the authors would like to express their appreciation for the support Junko Prasad provided in finishing the book.

1

Introduction

1.1 Mobile Communications Systems: Past, Present, and Future

There has been a paradigm shift in mobile communications systems every decade. The first generation (1G) systems in the 1980s were based on analog technologies, and the second generation (2G) systems in the 1990s, such as Global Systems for Mobile Telecommunications (GSM) [1], Personal Digital Cellular (PDC) [2], and Interim Standard (IS)-95 [3], on digital technologies for voice-oriented traffic. The 3G systems are also based on digital technologies for mixed voice, data, and multimedia traffic and mixed-circuit and packet-switched network [4, 5]. The first 3G system was introduced in October 2001 in Japan [6].

Figure 1.1 shows a rough sketch of present and future mobile communications systems. As an evolutional form of mobile phone systems, International Mobile Telecommunications (IMT)-2000 [4, 5], which corresponds to 3G systems, aims to support a wide range of multimedia services from voice and low-rate to high-rate data with up to at least 144 Kbps in vehicular, 384 Kbps in outdoor-to-indoor, and 2 Mbps in indoor and picocell environments. It provides continuous service coverage in 2-GHz band with code division multiplexing/code division multiple access (CDM/CDMA) scheme and supports both circuit-switched and packet-oriented services. Furthermore, high data rate (HDR), which supports a maximum 2.4-Mbps downlink packet transmission, is proposed [7], and high-speed downlink packet access (HSDPA), which also aims for more than 2-Mbps throughput, is under

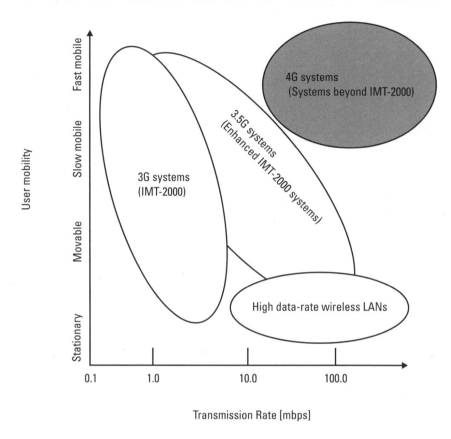

Figure 1.1 A rough sketch of present and future mobile communications systems.

standardization in the Third Generation Partnership Project (3GPP) [8]. Both HDR and HSDPA are categorized into enhanced IMT-2000 systems, which correspond to 3.5G systems.

As a progressive form of wireless local area networks (LANs), high-rate wireless LANs [9] such as IEEE802.11a [10], high-performance radio LAN type two (HIPERLAN/2) [11], and multimedia mobile access communication (MMAC) [12, 13], which are all based on the OFDM technique, provide data transmission up to 54 Mbps in 5-GHz band. They are mainly intended for communications between computers in an indoor environment, although they can support real-time audio and video transmission, and users are allowed some mobility [14].

1.2 Toward 4G Systems

Long-term researches and developments are usually required to lead a commercial service to success. Now, just coming into the new century, it might be a good time to start discussions on 4G systems, which may be put in service around 2010. Indeed, since the beginning of this century, we have often seen the words "future generation," "beyond 3G" or "4G" in magazines on wireless communications [15–19].

According to the Vision Preliminary Draft of New Recommendation (DNR) of ITU-R WP8F [20, 21], there will be a steady and continuous evolution of IMT-2000 to support new applications, products, and services. For example, the capacities of some of the IMT-2000 terrestrial radio interfaces are already being extended up to 10 Mbps, and it is anticipated that these will be extended even further, up to approximately 30 Mbps, by 2005, although these data rates will be limited only under optimum signal and traffic conditions. For systems beyond 3G [beyond IMT-2000 in the International Telecommunication Union (ITU)], there may be a requirement for a new wireless access technology for the terrestrial component around 2010. This will complement the enhanced IMT-2000 systems and the other radio systems with which there is an interrelationship. It is envisaged that these potential new radio interfaces will support up to approximately 100 Mbps for high mobility and up to approximately 1 Gbps for low mobility such as nomadic/local wireless access by around 2010.

The data rate figures are targets for research and investigation on the basic technologies necessary to implement the vision. The future system specification and design will be based on the results of the research and investigations. Due to the high data rate requirements, additional spectrum will be needed for these new capabilities of systems beyond IMT-2000. The data rate targets consider advances in technology, and these values are expected to be feasible from a technology perspective in the time frame of investigation and development of the new capabilities of systems beyond IMT-2000.

In conjunction with the future development of IMT-2000 and systems beyond IMT-2000, there will be an increasing relationship between radio access and communication systems, such as wireless personal area networks (PANs), LANs, digital broadcast, and fixed wireless access.

Based on today's envisaged service requirements, traffic expectations, and radio access technologies, ITU-R is working on a potential system architecture, according to Figures 1.2 through 1.4.

In this context, low mobility covers pedestrian speed ($\approx$3km/h), medium mobility corresponds to limited speed as for cars within cities ($\approx$50-60 km/h), high mobility covers high speed as on highways or with fast trains ($\approx$60 km/h to 250 km/h, or even more). The degree of mobility is basically linked to the cell size in a cellular system, as well as to system capacity. In general, the cell size in a cellular system has to be greater for a higher degree of mobility in order to limit the handover load in the network.

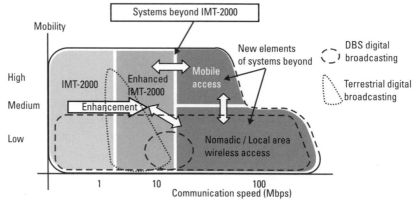

⟺ denotes interconnection between systems via networks or the like, which allows flexible use in any environments without making users aware of constituent systems.

Figure 1.2 System capabilities for systems beyond third generation. (*After:* [20].)

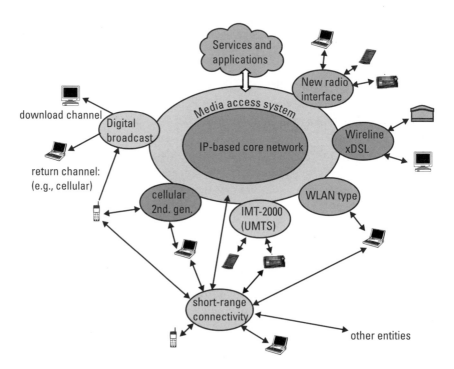

Figure 1.3 Seamless future network, including a variety of interworking access systems. (*After:* [20].)

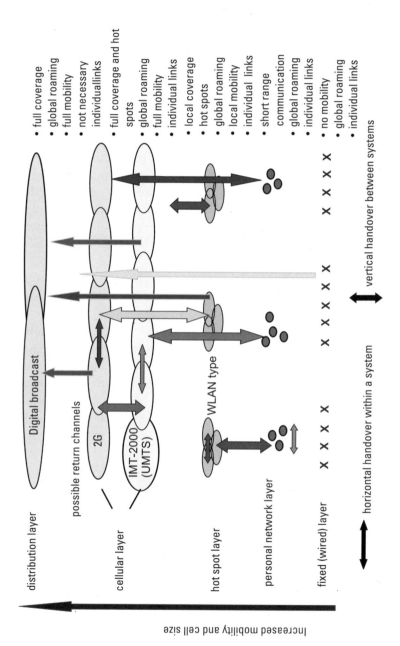

Figure 1.4 Layered structure of seamless future network with primarily allocated systems. (*After:* [20].)

Different complementary access schemes will be part of systems beyond IMT-2000. The different access systems are cooperating in terms of vertical handover and seamless service provision. Reconfigurable terminal devices and network infrastructure will be an essential part of such architecture. Such a concept of heterogeneous networks enables a migration and evolution path for network operators from today's networks to systems beyond IMT-2000 by reusing deployed investment. New access components can be added where and when needed from economic reasons. This ensures the requested scalability of the system according to Figure 1.5.

Possible new radio interface components are part of the concept. The different access systems will use already-allocated and identified frequency bands and potential new frequency bands for the new elements. Therefore, no direct interference between different technologies has to be expected. All access systems will be connected to an Internet Protocol (IP)-based network. Discussions are ongoing as to whether there should be a distinction between radio access and core network in the future.

From today's perspective, ITU-R expects the start of system standardization after WRC'07 with respect to identified spectrum bands and an initial deployment of systems beyond IMT-2000 after 2010.

The future system will comprise available and evolving access technologies. In addition, new radio access technologies with high carrier data rate for the wireless nomadic case with low mobility and for the cellular case with high mobility are envisaged. The data rate requirement for the cellular case is a big challenge from the technological perspective and with respect to the availability of sufficient future spectrum. Figure 1.6 shows the time plan for the systems beyond IMT-2000.

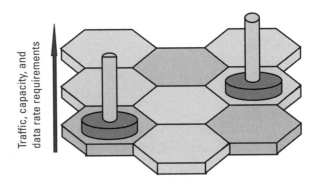

Figure 1.5 Inhomogeneous traffic or system capacity demand in deployment area. (*After:* [20].)

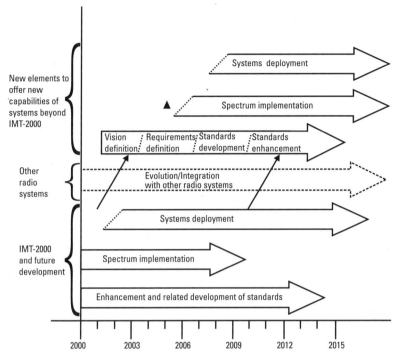

The dotted lines indicate that the exact starting point of the particular subject cannot yet be fixed.

▲ : Spectrum identification assumes the WRC03 approved WRC06 agenda and WRC06 identifies the spectrum.

Figure 1.6 Timelines. (*After:* [21].)

1.3 Multicarrier Techniques for 4G Systems

Figure 1.7 shows the evolution of mobile communications systems. In discussions about 2G systems in the 1980s, two candidates for the radio access technique existed, time division multiple access (TDMA) and CDMA schemes. Finally, the TDMA scheme was adopted as the standard. On the other hand, in the discussions about 3G systems in the 1990s, there were also two candidates, the CDMA scheme, which was adopted in the one-generation older systems, and the OFDM-based multiple access scheme called band division multiple access (BDMA) [22]. CDMA was finally adopted as the standard. If history is repeated, namely, if the radio access technique that was once not adopted can become a standard in new generation systems, then the OFDM-based technique looks promising as a 4G standard.

This book presents multicarrier techniques, including OFDM, and we believe that readers can easily understand the reason why it is suited for 4G

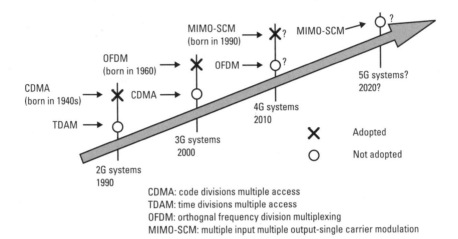

Figure 1.7 History of mobile communications systems in terms of adopted radio access technique.

systems when they finish reading. The following provides some of our justifications:

1. Multicarrier techniques can combat hostile frequency selective fading encountered in mobile communications. The robustness against frequency selective fading is very attractive, especially for high-speed data transmission [15].

2. OFDM scheme has been well matured through research and development for high-rate wireless LANs and terrestrial digital video broadcasting. We have developed a lot of know-how on OFDM.

3. By combining OFDM with CDMA, we can have synergistic effects, such as enhancement of robustness against frequency selective fading and high scalability in possible data transmission rate.

Figure 1.8 shows the advantages of multicarrier techniques.

1.4 Preview of the Book

This book is composed of 10 chapters. It covers all the necessary elements to understand multicarrier-based techniques. Chapter 2 briefly shows the characteristics of radio channels. It is the prerequisite knowledge for readers to understand the phenomena in the radio channels and is essential to carry

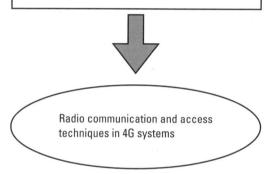

Figure 1.8 Advantages of multicarrier techniques for 4G systems.

out theoretical analysis and performance evaluation on multicarrier technique in radio channels.

Chapter 3 shows the history and principle of multicarrier technique, including the OFDM scheme. It includes the history from the origin to the current form.

Chapter 4 discusses the characteristics of the OFDM scheme. It puts much emphasis on the theoretical analysis and discusses advantages and disadvantages of OFDM, including robustness against frequency selective fading and impulsive noises and sensitivity to frequency offset and nonlinear amplification.

Synchronization of carrier frequency and discrete Fourier transform (DFT) window timing is a very important task for an OFDM receiver. Two chapters are devoted to this topic; Chapter 5 shows several pilot-assisted approaches on the synchronization, whereas Chapter 6 deals with a pilotless approach.

To obtain diversity effect in fading channels, channel coding/decoding scheme is essential. Chapter 7 shows the frequency diversity effect in the performance of coded OFDM scheme. It discusses symbol- and bit-interleaving depth to obtain a full frequency diversity effect.

Chapter 8 shows several systems where OFDM was successful. It includes digital audio broadcasting (DAB), terrestrial digital video broadcasting (DVB-T), terrestrial integrated services digital broadcasting (ISDB-T), IEEE 802.11a, HIPERLAN/2, MMAC, IEEE 802.11g, IEEE 802.11h, and IEEE 802.16a.

Chapter 9 discusses a combination of OFDM and CDMA. One combination is called multicarrier code division multiple access (MC-CDMA). Since it was born in 1993, intensive research has been conducted on this interesting new access scheme, and it is now considered suitable for a radio access technique in 4G systems. This chapter shows the principle and performance of the MC-CDMA.

Finally, Chapter 10 presents some recent (2000–2002) interesting research topics related to multicarrier technologies for future research.

References

[1] Mouly, M., and M. B. Pautet, "Current Evolution of the GSM," *IEEE Personal Commun. Mag.,* Vol. 2, No. 5, October 1995, pp. 9–19.

[2] Kinoshita, K., and M. Nakagawa, "Japanese Cellular Standard," *The Mobile Communications Handbook,* J. D. Gipson (ed.), Boca Raton, FL: CRC Press, pp. 449–461, 1996.

[3] Ross, A. H. M., and K. L. Gilhousen, "CDMA Technology and the IS-95 North American Standard," *The Mobile Communications Handbook,* J. D. Gipson (ed.), Boca Raton, FL: CRC Press, pp. 430–448, 1996.

[4] Prasad, R., *CDMA for Wireless Personal Communications,* Norwood, MA: Artech House, 1996.

[5] Ojanpera, T., and R. Prasad, (eds.), *Wideband CDMA for Third Generation Mobile Communications,* Norwood, MA: Artech House, 1998.

[6] Ohmori, S., Y. Yamao, and N. Nakajima, "The Future Generations of Mobile Communications Based on Broadband Access Technologies," *IEEE Commun. Mag.,* Vol. 38, No. 12, December 2000, pp.134–142.

[7] Bender, P., et al., "CDMA/HDR: A Bandwidth-Efficient High-Speed Wireless Data Services for Nomadic Users," *IEEE Commun. Mag.,* Vol. 38, No. 7, July 2000, pp. 70–77.

[8] 3 GPP, 3G TR25.848, V.0.6.0, May 2000.

[9] van Nee, R., et al., "New High-Rate Wireless LAN Standards," *IEEE Commun. Mag.,* Vol. 37, No. 12, December 1999, pp. 82–88.

[10] IEEE Std. 802.11a, "Wireless Medium Access Control (MAC) and Physical Layer (PHY) Specifications: High-speed Physical Layer Extension in the 5-GHz Band," IEEE, 1999.

[11] ETSI TR 101 475, "Broadband Radio Access Networks (BRAN); HIPERLAN Type2; Physical (PHY) Layer," ETSI BRAN, 2000.

[12] ARIB STD-T70, "Lower Power Data Communication Systems Broadband Mobile Access Communication System (CSMA)," ARIB, December 2000.

[13] ARIB STD-T70, "Lower Power Data Communication Systems Broadband Mobile Access Communication System (HiSWANa)," ARIB, December 2000.

[14] van Nee, R., and R. Prasad, *OFDM for Wireless Multimedia Communications,* Norwood, MA: Artech House, 2000.

[15] Chuang, J., and N. Sollenberger, "Beyond 3G: Wideband Wireless Data Access Based on OFDM and Dynamic Packet Assignment," *IEEE Commun. Mag.,* Vol. 38, No. 7, July 2000, pp. 78–87.

[16] "Fourth Generation Wireless Networks and Interconnecting Standards," *IEEE Personal Commun. Mag.,* (special issue), Vol. 8, No. 5, October 2001.

[17] "European R&D on Fourth-Generation Mobile and Wireless IP Networks," *IEEE Personal Commun. Mag.,* (special issue), Vol. 8, No. 6, December 2001.

[18] "Mobile Initiatives and Technologies," *IEEE Commun. Mag.,* (special issue), Vol. 40, No. 3, March 2002.

[19] "Technologies for 4G Mobile," *IEEE Wireless Commun.,* Vol. 9, No. 2, April 2002.

[20] Mohr, W., "Heterogeneous Networks to Support User Needs with Major Challenges for New Wideband Access Systems," *Wireless Personal Communications* (Kluwer), Vol. 22, No. 2, August 2002, pp. 109–137.

[21] Nakagawa, H., "Vision in WP8F and Activity in Japan for Future Mobile Communication System," *Strategic Workshop 2002 Unified Global Infrastructure,* Prague, Czech Republic, September 6–7, 2002.

[22] ARIB FPLMTS Study Committee, "Report on FPLMTS Radio Transmission Technology Special Group (Round 2 activity report)," Draft v.E1.1, January 1997.

2

Characteristics of Multipath Fading Channels

2.1 Introduction

Radio propagation characterization is the bread and butter of communications engineers. Without knowledge of radio propagation, a wireless system could never be developed. Radio engineers have to acquire full knowledge of the channel if they want to be successful in designing a good radio communication system [1]. Therefore, knowledge of radio propagation characteristics is a prerequisite for designing radio communication systems.

A lot of measurements have been done to obtain information concerning multipath fading channels. Reference [2] has presented a good overview of this topic. Detailed discussions on characteristics of multipath fading channels can be found in [3–8]. This chapter shows the essence in the literature.

This chapter is organized as follows. Multipath fading is due to multipath reflections of a transmitted wave by local scatterers such as houses, buildings, and man-made structures, or natural objects such as forest surrounding a mobile unit. The probability density function of the received signal follows a Rayleigh or Ricean distribution. Section 2.2 presents the Rayleigh and Ricean fading channels, and multipath delay profile is discussed in Section 2.3. The wireless channel is defined as a link between a transmitter and a receiver and classified considering the coherence bandwidth and coherence time. Accordingly, a wireless channel can be frequency selective or frequency nonselective (explained in Section 2.4) and time selective or time

13

nonselective (described in Section 2.6). Section 2.5 briefly introduces the spaced-time correlation function.

2.2 Rayleigh and Ricean Fading Channels

Figure 2.1 shows a typical multipath fading channel often encountered in wireless communications, where there are L paths. Assume the transmitted signal is given by

$$x(t) = \mathrm{Re}\,[s(t)\,e^{j2\pi f_C t}] \tag{2.1}$$

where $s(t)$ is the equivalent baseband form of $x(t)$ and f_C is a carrier frequency. In addition, Re [*] denotes the real part of * (Im [*] denotes the imaginary part of *).

Through the multipath fading channel, the received signal is written as

$$y(t) = \sum_{l=1}^{L} \alpha_l(t)\,x(t - \tau_l(t)) \tag{2.2}$$

$$= \mathrm{Re}\left[\sum_{l=1}^{L} \alpha_l(t)\,e^{-j2\pi f_C \tau_l(t)}\,s(t - \tau_l(t))\,e^{j2\pi f_C t}\right]$$

where $\alpha_l(t)$ and $\tau_l(t)$ are the complex-valued channel loss (or gain) and real-valued time delay for the lth path, both of which can be modeled as

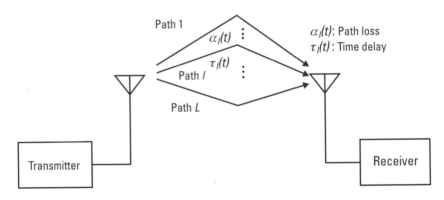

Figure 2.1 Example of a multipath fading channel.

stochastic processes. Then, the equivalent baseband form of $y(t)$ is written as

$$r(t) = \sum_{l=1}^{L} \alpha_l(t) e^{-j2\pi f_C \tau_l(t)} s(t - \tau_l(t)) \qquad (2.3)$$

$$= \int_{-\infty}^{+\infty} h(\tau; t) s(t - \tau) \, d\tau$$

where $h(\tau; t)$ is the equivalent baseband impulse response of the multipath fading channel at instant t, which is given by

$$h(\tau; t) = \sum_{l=1}^{L} \alpha_l(t) e^{-j2\pi f_C \tau_l(t)} \delta(t - \tau_l(t)) \qquad (2.4)$$

Assume that the transmitted signal is a continuous wave (CW) with frequency of f_C. In this case, if setting $s(t) = 1$ in (2.3), the received signal is written as

$$r(t) = \sum_{l=1}^{L} \alpha_l(t) e^{-j2\pi f_C \tau_l(t)} \qquad (2.5)$$

$$= \sum_{l=1}^{L} \beta_l(t)$$

$$\beta_l(t) = \alpha_l(t) e^{-j2\pi f_C \tau_l(t)} \qquad (2.6)$$

where $\beta_l(t)$ is a complex-valued stochastic process.

Equation (2.5) clearly shows that the received signal is the sum of stochastic processes, so when there are a large number of paths, the central limiting theorem can be applied. That is, $r(t)$ can be modeled as a complex-valued Gaussian stochastic process with its average and variance given by

$$av_r = E[r(t)], \quad \sigma_r^2 = \frac{1}{2} E[r^*(t) r(t)] \qquad (2.7)$$

therefore, the probability density function (p.d.f.) of $r = r(t)$ is given by

$$p(r) = \frac{1}{2\pi\sigma_r^2} e^{-\frac{(r - av_r)^*(r - av_r)}{2\sigma_r^2}} \qquad (2.8)$$

Furthermore, defining the envelope and phase of $r(t)$ as

$$\xi(t) = |r(t)|, \ \theta(t) = \arg r(t) \qquad (2.9)$$

the joint p.d.f. of $\xi = \xi(t)$ and $\theta = \theta(t)$ is given by

$$p(\xi, \theta) = \frac{\xi}{2\pi\sigma_r^2} e^{-\frac{1}{2\sigma_r^2}[(\xi \cos \theta - a_I)^2 + (\xi \sin \theta - a_Q)^2]} \qquad (2.10)$$

$$= \frac{\xi}{2\pi\sigma_r^2} e^{-\frac{\xi^2 + A^2}{2\sigma_r^2}} e^{\frac{\xi(a_I \cos \theta + a_Q \sin \theta)}{2\sigma_r^2}}$$

where

$$a_I = \text{Re}[av_r], \ a_Q = \text{Im}[av_r] \qquad (2.11)$$

$$A = |av_r| \qquad (2.12)$$

Averaging (2.10) in terms of θ, the p.d.f. of ξ is given by

$$p(\xi) = \int_0^{2\pi} p(\xi, \theta) \, d\theta \qquad (2.13)$$

$$= \frac{\xi}{\sigma_r^2} I_0 \left(\frac{\xi A}{\sigma_r^2}\right) e^{-\frac{\xi^2 + A^2}{2\sigma_r^2}}, \ (\xi > 0)$$

where $I_0(x)$ is the *zero*th order modified Bessel function of the first kind, which is defined as

$$I_0(x) = \frac{1}{2\pi} \int_0^{2\pi} e^{x \cos \theta} \, d\theta \qquad (2.14)$$

The p.d.f. of the envelope given by (2.13) is called the Rice distribution [3]. Especially, in (2.13),

$$K = \frac{A^2}{2\sigma_r^2} \tag{2.15}$$

is called "the Ricean K factor."

When $r(t)$ can be modeled as a zero average complex-valued Gaussian stochastic process, that is, $A = 0$ in (2.13), the p.d.f. of ξ is given by

$$p(\xi) = \frac{\xi}{\sigma_r^2} e^{-\frac{\xi^2}{2\sigma_r^2}}, \ (\xi > 0) \tag{2.16}$$

In this case, the p.d.f. of the phase is uniformly distributed as

$$p(\theta) = \frac{1}{2\pi}, \ (0 < \theta \leq 2\pi) \tag{2.17}$$

The p.d.f. of the envelope given by (2.16) is called the Rayleigh distribution [3]. Note that $p(\xi)$ and $p(\theta)$ are statistically independent. Figure 2.2 shows the Rice and Rayleigh distributions.

The multipath propagation model for the received signal $r(t)$, given by (2.3), results in signal fading. The previous discussion shows that, when the impulse response is modeled as a zero average complex-valued Gaussian process, the envelope at any instant t is Rayleigh-distributed. The Rayleigh

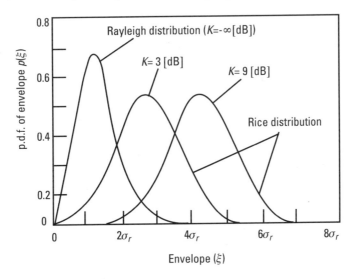

Figure 2.2 Rice and Rayleigh distributions.

distribution is commonly used to describe the statistical time varying nature of the envelope of a frequency nonselective (flat) fading signal, or the envelope of an individual multipath component. In this case, the channel is called "a Rayleigh fading channel."

On the other hand, when a direct path is available or the channel has signal reflectors, $r(t)$ cannot be modeled as a zero average process. In this case, the envelope has a Rice distribution, and the channel is called "a Ricean fading channel."

From Figure 2.2, we can see that the Rayleigh fading is a kind of a deep fading, as compared with Ricean fading. This is clear from the fact that the Rayleigh distribution tends to have smaller values in the envelope. In addition, from the definition of the Ricean K factor given by (2.15), as $K \to -\infty$ dB ($A \to 0$), the Ricean distribution degenerates to a Rayleigh distribution.

2.3 Multipath Delay Profile

Assuming that $h(\tau; t)$ is wide sense stationary (WSS), its autocorrelation function is given by [8]

$$\phi_h(\tau_1, \tau_2; \Delta t) = \frac{1}{2} E[h^*(\tau_1; t) h(\tau_2; t + \Delta t)] \qquad (2.18)$$

Furthermore, assuming that the loss and phase shift of the channel associated with path delay τ_1 is uncorrelated with the loss and phase shift of the channel associated with path delay τ_2 [this is called uncorrelated scattering (US)], we obtain

$$\phi_h(\tau_1, \tau_2; \Delta t) = \phi_h(\tau_1; \Delta t) \delta(\tau_1 - \tau_2) \qquad (2.19)$$

where $\delta(\tau)$ is the Dirac's Delta function. When setting $\Delta t = 0$, $\phi_h(\tau) = \phi_h(\tau; 0)$ is called the multipath delay profile or multipath intensity profile, describing the *average* power output of the channel as a function of the time delay τ:

$$\phi_h(\tau) = \phi_h(\tau; 0) = \frac{1}{2} E[h^*(\tau; t) h(\tau; t)] \qquad (2.20)$$

Figure 2.3 shows the relation between $\phi_h(\tau)$ and $h(\tau; t)$.

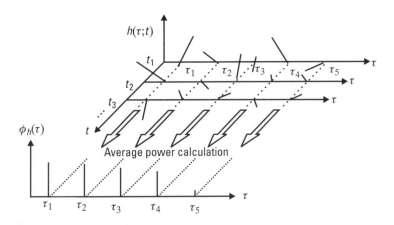

Figure 2.3 Relation between $\phi_h(\tau)$ and $h(\tau; t)$.

2.4 Frequency Selective and Frequency Nonselective Fading Channels

The equivalent baseband form of the transfer function for the channel at instant t is obtained from the Fourier transform of $h(\tau; t)$:

$$H(f; t) = \int_{-\infty}^{+\infty} h(\tau; t)\, e^{-j2\pi f\tau}\, d\tau \qquad (2.21)$$

where f denotes the frequency. When $h(\tau; t)$ is a WSSUS Gaussian stochastic process, $H(f; t)$, which is given by the Fourier transform of $h(\tau; t)$, is also a WSSUS Gaussian stochastic process, so its autocorrelation function can be defined as

$$\phi_H(\Delta f; \Delta t) = \frac{1}{2} E[H^*(f; t) H(f + \Delta f; t + \Delta t)] \qquad (2.22)$$

Substituting (2.20) and (2.21) into (2.22) leads to:

$$\phi_H(\Delta f; \Delta t) = \int_{-\infty}^{+\infty} \phi_h(\tau; \Delta t)\, e^{-j2\pi\tau\Delta f}\, d\tau \qquad (2.23)$$

In (2.23), if setting $\Delta t = 0$, we obtain the spaced-frequency correlation function of the channel $\phi_H(\Delta f) = \phi_H(\Delta f; 0)$

$$\phi_H(\Delta f) = \int\limits_{-\infty}^{+\infty} \phi_h(\tau) e^{-j2\pi\Delta f\tau} \, d\tau \qquad (2.24)$$

which describes the correlation between frequency variations of the channel separated by Δf.

The multipath channel generally has a bandwidth where channel variations are highly correlated, that is, $\phi_H(\Delta f)/\phi_H(0)$ can be approximated as 1.0. This bandwidth is called the coherence bandwidth $(\Delta f)_c$. When a signal is transmitted through a channel, if $(\Delta f)_c$ of the channel is small compared with the bandwidth of the transmitted signal, the channel is called to be *frequency selective*. In this case, the signal is severely distorted by the channel. On the other hand, if $(\Delta f)_c$ is much larger compared with the bandwidth of the transmitted signal, the channel is called to be *frequency nonselective* or *flat*.

Similar to the coherence bandwidth, as a measure for frequency selectivity of the channel, there are two important parameters, the average excess delay and the root mean square (RMS) delay spread. They are respectively defined as

$$T_{AEX} = \frac{1}{\sigma_r^2} \int\limits_{-\infty}^{+\infty} \tau\phi_h(\tau) \, d\tau \qquad (2.25)$$

$$\tau_{RMS} = \sqrt{\frac{1}{\sigma_r^2} \int\limits_{-\infty}^{+\infty} \tau^2\phi_h(\tau) \, d\tau - T_{AEX}^2} \qquad (2.26)$$

2.5 Spaced-Time Correlation Function

In (2.23), if setting $\Delta f = 0$, we obtain the spaced-time correlation function of the channel $\phi_H(\Delta t) = \phi_H(0; \Delta t)$, describing the correlation between time variations of the channel separated by Δt. The Doppler power spectrum is defined as its Fourier transform:

$$D_H(\varsigma) = \int\limits_{-\infty}^{+\infty} \phi_H(\Delta t) e^{-j2\pi\varsigma\Delta t} \, d\Delta t \qquad (2.27)$$

$$\phi_H(\Delta t) = \int\limits_{-\infty}^{+\infty} D_H(\varsigma) e^{j2\pi\varsigma\Delta t} \, d\varsigma \qquad (2.28)$$

Figure 2.4 shows the relation between $H(f; t)$ and $D_H(\varsigma)$.

2.6 Time Selective and Time Nonselective Fading Channels

The multipath channel generally also has a time duration where channel variations are highly correlated, that is, $\phi_H(\Delta t)/\phi_H(0)$ can be approximated as 1.0. This time duration is called the coherence time $(\Delta t)_c$. When a signal is transmitted through a channel, if $(\Delta t)_c$ of the channel is small compared with the symbol duration of the transmitted signal, the channel is called to be *time selective* or *fast*. On the other hand, if $(\Delta t)_c$ is much larger compared with the symbol duration of the transmitted signal, the channel is called to be *time nonselective* or *slow*.

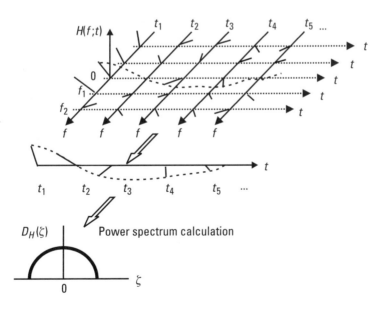

Figure 2.4 Relation between $H(f; t)$ and $D_H(\varsigma)$.

2.7 Examples of Multipath Fading Channels

As discussed, when a signal is transmitted through a multipath fading channel, a channel model, which we assume, gives its received characteristics. For instance, the p.d.f. such as Rayleigh distribution and Ricean distribution describes the envelope fluctuation for an individual multipath component in the channel, the multipath intensity profile or spaced-frequency correlation function determines the frequency selectivity of the channel, and the Doppler power spectrum or spaced-time correlation function determines the time selectivity of the channel.

Figure 2.5 shows examples of multipath delay profiles to describe frequency selectivity of a channel, which we often use in this book for computer simulation. Here, we can assume that the fading characteristic of each path is independent because of WSSUS, and we often assume a Rayleigh distribution for each envelope and a uniform distribution for each phase. In Figure 2.5(a), there are a fixed number of paths with equidistant delays and the average received powers of multipaths are exponentially decaying. We often encounter this kind of profile in indoor environments [9, 10] and we call it "an exponentially decaying profile." On the other hand, in Figure 2.5(b), there are also a fixed number of paths with equidistant delays but the average received powers of multipaths are all the same. We often use this kind of profile to test a system performance [6] and we call it "an independent and identically distributed (i.i.d.) profile."

The time variation of a channel is determined by a lot of factors, such as the height of transmitter/receiver antenna, the speed of transmitter/receiver

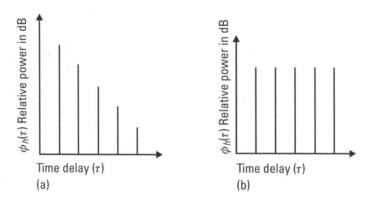

Figure 2.5 Examples of multipath delay profiles: (a) an exponentially decaying multipath delay profile; and (b) an i.i.d. multipath delay profile.

in motion, the shape of the antenna, the height of surrounding structures, and so on.

Figure 2.6 shows a situation where a receiver with an omnidirectional antenna is in motion with velocity of v and a lot of signals arrive at the antenna in all directions. This model is called the "Jakes' model," [4, 5] and in this case, defining the direction arrival of a signal from the direction of motion as θ, the Doppler shift for the signal is given by

$$\varsigma = \frac{v}{\lambda} \cos \theta = f_D \cos \theta \qquad (2.29)$$

where λ is the wavelength and f_D is the maximum Doppler shift. Defining the power spectrum density as $D_H(\zeta)$, the power of the received signal in frequency range of $[\zeta,\ \zeta + d\zeta]$ is given by

$$D_H(\varsigma)|d\varsigma| = 2 \times \frac{\sigma_r^2}{2\pi}|d\theta| \qquad (2.30)$$

Differentiating (2.28) leads to:

$$d\varsigma = -f_D \sin \theta \, d\theta \qquad (2.31)$$

therefore, finally, substituting (2.31) into (2.30), we obtain the Doppler power spectrum:

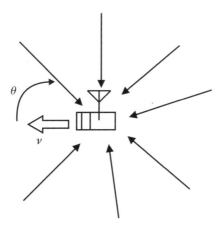

Figure 2.6 Jakes' model.

$$D_H(s) = \frac{\sigma_r^2}{\pi\sqrt{f_D^2 - s^2}} \tag{2.32}$$

Figure 2.7 shows the Doppler power spectrum. Equation (2.32) means that the bandwidth of the received signal is broadened; in other words, the signal is randomly frequency-modulated by noise through the channel. We call this noise "random FM noise."

Substituting (2.32) into (2.28) leads to:

$$\phi_H(\Delta t) = \sigma_r^2 J_0(2\pi f_D \Delta t) \tag{2.33}$$

where $J_0(x)$ is the *zero*th Bessel function of the first kind.

Figure 2.8 shows a time variation of signal through a frequency nonselective Rayleigh fading channel. The envelope and phase (see Figure 2.8(a, b), respectively) are obtained by computer simulation based on the following equation [11]:

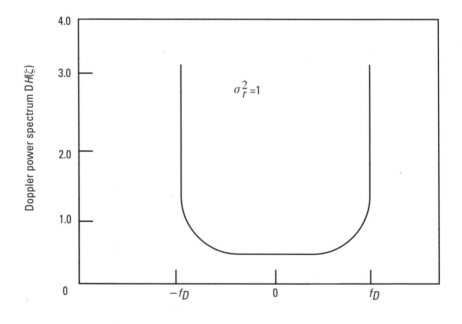

Frequency (ζ) [Hz]

Figure 2.7 Doppler power spectrum.

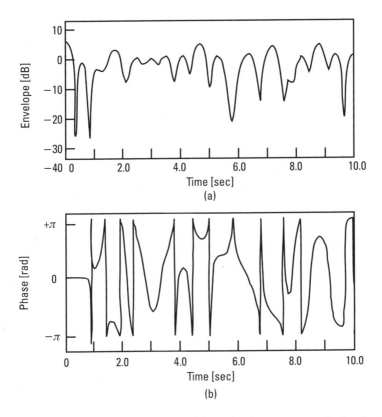

Figure 2.8 Time variation of a signal received through a frequency nonselective Rayleigh fading channel: (a) envelope; and (b) phase.

$$r(t) = \sum_{l=1}^{L} e^{j2\pi f_D \cos\left(\frac{2\pi l}{L}\right) t} \qquad (2.34)$$

where we used $L = 9$ and $f_D = 1$ Hz. It is interesting to note that (2.34) means a sum of deterministic signals with the same amplitude, but because of the central limit theorem, the envelope and phase of the resultant signal are Rayleigh and uniformly distributed, respectively.

The three factors to describe the fading characteristics that a transmitted signal experiences in a channel, such as the p.d.f. of the envelope, frequency selectivity, and time selectivity, are independent, so there are many combinations to consider. For instance, when no line-of-sight component is available in a channel, the data transmission rate is very high, and the receiver is installed in a high-speed cruising vehicle, the channel will be "a frequency selective fast Rayleigh fading channel," whereas when a line-of-sight component is available in a channel, the data transmission rate is very low and the receiver is installed in a stationary terminal, the channel will be "a frequency nonselective slow Ricean fading channel."

References

[1]   Prasad, R., "European Radio Propagation and Subsystems Research for the Evolution of Mobile Communications," *IEEE Comm. Mag.*, Vol. 34, No. 2, February 1996, p. 58.

[2]   Prasad, R., *Universal Wireless Personal Communications*, Norwood, MA: Artech House, 1998.

[3]   Schwartz, M., W. R. Bennett, and S. Stein, *Communication Systems and Techniques*, New York: IEEE Press, 1996.

[4]   Jakes, Jr., C., *Microwave Mobile Communications*, New York: John Wiley & Sons, 1974.

[5]   Lee, W. C. Y., *Mobile Communications Engineering*, New York: McGraw-Hill, 1982.

[6]   Steele, R., *Mobile Radio Communications*, New York: IEEE Press, 1992.

[7]   Rappaport, T. S., *Wireless Communications*, Piscataway, NJ: Prentice Hall, 1996.

[8]   Proakis, J. G., *Digital Communications, Fourth Edition*, New York: McGraw-Hill, 2001.

[9]   Saleh, A. A., and R. A. Valenzuela, "A Statistical Model for Indoor Multipath Propagation," *IEEE J. Select. Areas Commun.*, Vol. SAC-5, No. 2, February 1987, pp. 128–137.

[10]   Hashemi, H., "The Indoor Radio Propagation Channel," *Proc. IEEE*, Vol. 81, No. 7, July 1993, pp. 943–968.

[11]   Cavers, J. K., *Mobile Channel Characteristics*, Boston, MA: Kluwer Academic Publishers, 2000.

3

Principle and History of MCM/OFDM

3.1 Introduction

OFDM is a special form of multicarrier modulation (MCM), where a single data stream is transmitted over a number of lower rate subcarriers. It is worth mentioning here that OFDM can be seen as either a modulation technique or a multiplexing technique.

One of the main reasons to use OFDM is to increase the robustness against frequency selective fading and narrowband interference. In a single carrier system, a single fade or interferer can cause the entire link to fail, but in a multicarrier system, only a small percentage of subcarriers will be affected. Error correction coding can then be used to correct the few erroneous subcarriers.

An introduction to OFDM is given in [1], and [2] presents the basic concept of OFDM and its applications. This chapter revisits the principle of OFDM and its history from its origin to current form. Throughput this chapter, we discuss the details in equivalent baseband expression.

This chapter is organized as follows. The origin of OFDM is discussed in Section 3.2, Section 3.3 presents the use of discrete Fourier transform (DFT), and the insertion of cyclic prefix for OFDM's current form is explained in Section 3.4. Section 3.5 concludes the topic.

3.2 Origin of OFDM

MCM is the principle of transmitting data by dividing input stream into several symbol streams, each of which has a much lower symbol rate, and by using these substreams to modulate several subcarriers.

Figure 3.1 compares a singlecarrier modulation (SCM) and an MCM. Here, B_{SCM} and B_{MCM} denote the bandwidths of transmitted SCM and MCM signals, respectively. For MCM, f_k, $F_k(f; t)$, N_{SC} and Δf denote the frequency of the kth subcarrier, the frequency spectrum of pulse waveform of the kth subcarrier, the total number of subcarriers, and subcarrier separation, respectively. The frequency spectrum of the MCM signal is written as

$$S_{MCM}(f; t) = \sum_{k=1}^{N_{SC}} F_k(f; t) \qquad (3.1)$$

Through a frequency selective fading channel characterized by the transfer function $H(f; t)$, the frequency spectra of received SCM and MCM signals are written as

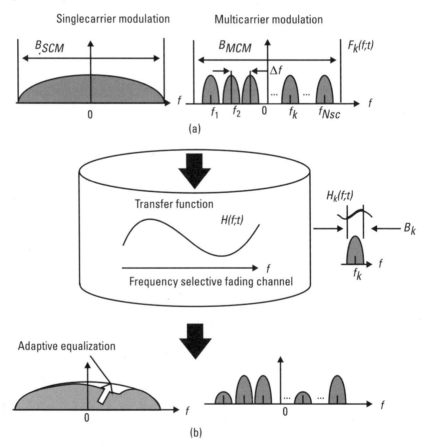

(a)

(b)

Figure 3.1 Comparison of SCM and MCM: (a) frequency spectra of transmitted signals; and (b) frequency spectra of received signals.

$$R_{SCM}(f; t) = H(f; t) S_{SCM}(f; t)$$

$$R_{MCM}(f; t) = H(f; t) S_{MCM}(f; t) \tag{3.2}$$

$$= \sum_{k=1}^{N_{SC}} H_k(f; t) F_k(f; t) \tag{3.3}$$

where $S_{SCM}(f; t)$ is the frequency spectrum of the transmitted SCM signal and $H_k(f; t)$ is the channel transfer function corresponding to B_k, which is the frequency band occupied by the kth subcarrier. When the number of subcarriers is large, the amplitude and phase response of $H_k(f; t)$ can be assumed to be constant over B_k, so $R_{MCM}(f; t)$ can be approximated as

$$R_{MCM}(f; t) \cong \sum_{k=1}^{N_{SC}} H_k(t) F_k(f; t) \tag{3.4}$$

where $H_k(t)$ is the complex-valued pass loss for B_k.

Equation (3.4) clearly shows that MCM is effective and robust in wireless channels; namely, to combat frequency selective fading, MCM requires no equalization or at most one-tap equalization for each subcarrier, whereas SCM requires complicated adaptive equalization.

The first systems employing MCM were military HF radio links in the late 1950s and early 1960s, such as KINEPLEX [3] and KATHRYN [4], where nonoverlapped band-limited orthogonal signals were used because of the difficulty in the precise control of frequencies of subcarrier local oscillators and the detection of subcarrier signals with analog filters.

According to the authors' best knowledge, the concept of MCM scheme employing time-limited orthogonal signals, which is all the same as OFDM, dates back to 1960 [5].

Defining the symbol duration at subcarrier level as T_s, the transmitted signal $s(t)$ is written as

$$s(t) = \sum_{i=-\infty}^{+\infty} \sum_{k=1}^{N_{SC}} c_{ki} e^{j2\pi f_k(t - iT_s)} f(t - iT_s) \tag{3.5}$$

where c_{ki} is the ith information symbol at the kth subcarrier, and $f(t)$ is the pulse waveform of the symbol. When the rectangular pulse waveform is used, $f(t)$ is given by

$$f(t) = \begin{cases} 1, & (0 < t \leq T_s) \\ 0, & (t \leq 0, \, t > T_s) \end{cases} \qquad (3.6)$$

so f_k and Δf are respectively written as

$$f_k = \frac{(k-1)}{T_s}, \; \Delta f = \frac{1}{T_s} \qquad (3.7)$$

Figure 3.2 compares a baseband serial data transmission system with an OFDM system. Here, the subscript i is dropped in (3.5) for the sake of simplicity.

Figure 3.3 compares the frequency spectra. The classical MCM, employing nonoverlapped band-limited orthogonal signals, matches the use of analog subcarrier oscillators and filters, but it requires much wider bandwidth [see Figure 3.3(a)]. If employing the rectangular pulse waveforms for subcarriers, the frequency spectra of the waveforms are widely spread and overlapped [see Figure 3.3(b)], although it can save the required bandwidth. Therefore, the concept of OFDM was once abandoned because of difficulty in subcarrier recovery without intersubcarrier interference by means of analog filters. As a result, a number of studies in the 1960s were dedicated for MCM employing overlapped band-limited orthogonal signals [see Figure 3.3(c)]. This is because analog filters can easily separate such signals [6, 7]. Note that the name of "OFDM" appeared in the U.S. Patent No. 3 issued in 1970 [8].

Figure 3.3 also contains the frequency spectrum of a time-limited signal [see Figure 3.3(d)]. Figures 3.3(b) and (d) correspond to Figures 3.2(b) and (a), respectively.

3.3 Use of Discrete Fourier Transform

The idea of using the DFT revived the MCM employing time-limited orthogonal signals, namely, OFDM [9].

Sampling $s(t)$ $(iT_s < t \leq (i + 1)T_s)$ with sampling rate of t_{spl} $(= T_s/N_{SC})$, the transmitted signal is written in a column vector form as

$$\mathbf{S} = [s(iT_s + t_{spl}), \ldots, s(iT_s + qt_{spl}), \ldots, s(iT_s + N_{SC}\,t_{spl})]^T \qquad (3.8)$$

$$= \mathbf{W}^{-1}(N_{SC})\,\mathbf{C}_i$$

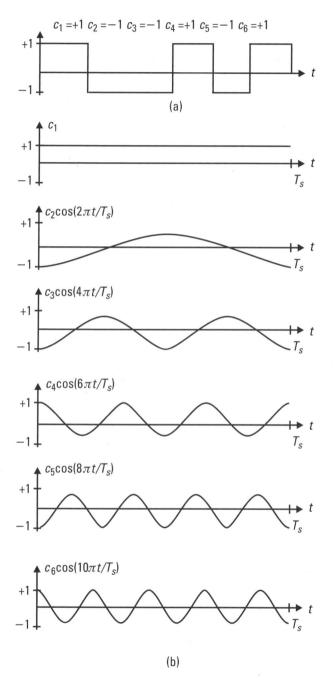

Figure 3.2 Comparison of transmitted waveforms: (a) Baseband serial data transmission system; and (b) OFDM system.

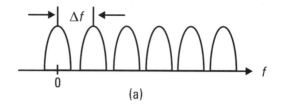

(a)

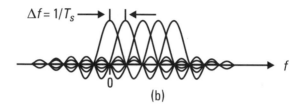

(b)

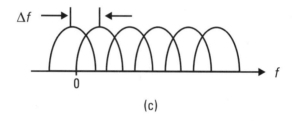

(c)

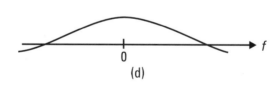

(d)

Figure 3.3 Comparison of frequency spectra: (a) nonoverlapped band-limited orthogonal signals (MCM); (b) overlapped time-limited orthogonal signals (MCM); (c) overlapped band-limited orthogonal signals (MCM); and (d) a time-limited signal (SCM).

where $^{*\mathrm{T}}$ denotes the transpose of $*$, and $\mathbf{W}^{-1}(N_{SC})$ and $\mathbf{C}_i$ are the N_{SC}-point inverse DFT (IDFT) matrix and the ith symbol (column) vector, respectively:

$$\mathbf{W}^{-1}(N_{SC}) = \{w_{qk}^{-1}\}$$

$$w_{qk}^{-1} = e^{j2\pi \frac{q(k-1)}{N_{SC}}} \qquad (3.9)$$

$$\mathbf{C}_i = [c_{1i}, c_{2i}, \ldots, c_{N_{SC}i}]^{\mathrm{T}} \qquad (3.10)$$

Equation (3.8) clearly shows that the transmitted symbol vector is recovered at the receiver by means of the DFT:

$$\mathbf{C}_i = \mathbf{W}(N_{SC})\mathbf{S} \qquad (3.11)$$

where $\mathbf{W}(N_{SC})$ is the N_{SC}-point DFT matrix given by

$$\mathbf{W}(N_{SC}) = \{w_{qk}\} \qquad (3.12)$$

$$w_{qk} = e^{-j2\pi \frac{q(k-1)}{N_{SC}}}$$

Figure 3.4 shows the OFDM system employing the IDFT and DFT. When employing rectangular DFT window at the receiver, intersubcarrier interference can be perfectly eliminated.

The use of IDFT/DFT totally eliminates bank of subcarrier oscillators at the transmitter/receiver, and furthermore, if selecting the number of subcarriers as the power of two, we can replace the DFT by the fast Fourier transform (FFT). The advantage of OFDM in mobile communications was first suggested in [10].

3.4 Insertion of Cyclic Prefix for Current Form of OFDM

Now, let us consider a distortion that a frequency selective fading channel gives to an OFDM signal. As shown in Chapter 2, frequency selective fading channel can be characterized by an impulse response with delay spread in the time domain, which is not negligibly small as compared with one symbol period. Figure 3.5 shows an instantaneous impulse response of a frequency selective fading channel, where we can see two paths and $\tau_{\max}$ denotes the time delay between the first and second paths.

Through the channel, the first path generates the desired signal and the second path the delayed signal at the receiver. Figure 3.6(a–c) shows three transmitted signals and Figure 3.7(a–c) shows the corresponding three

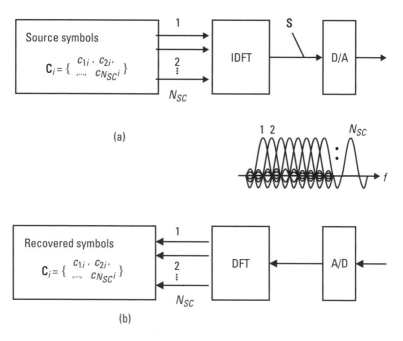

Figure 3.4 OFDM system: (a) transmitter; and (b) receiver.

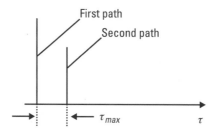

Figure 3.5 Instantaneous impulse response of a frequency selective fading channel.

received signals. Here, we pay attention only to waveforms at a certain subcarrier ($k = 2$).

Without a guard interval between successive OFDM symbols, intersymbol interference (ISI) from the $(i - 1)$th symbol gives a distortion to the ith symbol [compare Figure 3.6(a) with Figure 3.7(a) and see the thick line in Figure 3.7(a)]. If we employ a guard interval (no signal transmission) with length of $\Delta_G > \tau_{max}$, we can perfectly eliminate ISI, but a sudden change of waveform contains higher spectral components, so they result in intersubcarrier interference [compare Figure 3.6(b) with Figure 3.7(b) and see the thick line in Figure 3.7(b)].

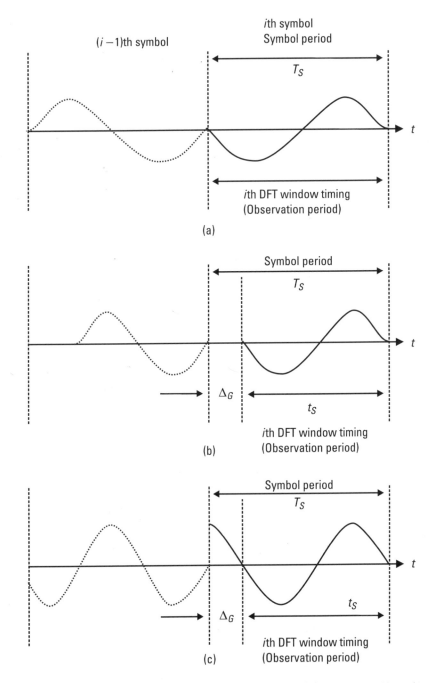

Figure 3.6 Transmitted signals: (a) no guard interval insertion; (b) guard interval insertion; and (c) guard interval insertion with cyclic prefix.

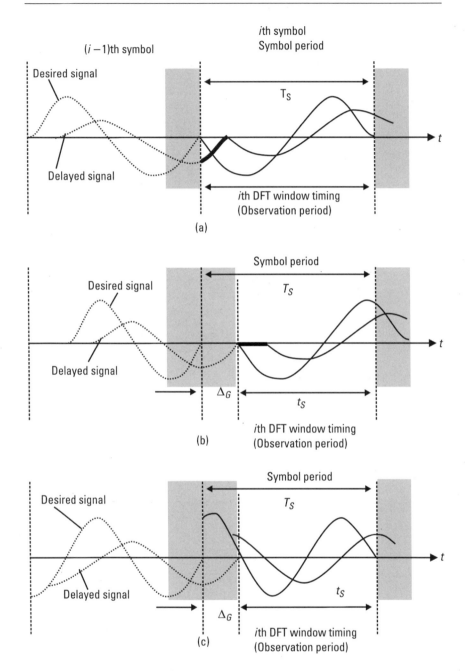

Figure 3.7 Received signals: (a) no guard interval insertion; (b) guard interval insertion; and (c) guard interval insertion with cyclic prefix.

Figure 3.6(c) shows the guard interval insertion technique with cyclic prefix to perfectly eliminate intersubcarrier interference, where the OFDM symbol is cyclically extended in the guard time [11, 12]. Paying attention to the ith DFT window with width of t_s in Figure 3.7(c), we can see two sinusoidal signals with full width, so it results in no intersubcarrier interference. Note that the OFDM symbol is T_s long but the subcarrier frequency is an integer multiple of $1/t_s$. This implies that the subcarrier separation now becomes a bit larger, namely, $1/t_s$.

Modifying (3.5), (3.6), and (3.7), the transmitted signal with the cyclic extension is finally written as

$$s(t) = \sum_{i=-\infty}^{+\infty} \sum_{k=1}^{N_{SC}} c_{ki} e^{j2\pi f_k (t - iT_s)} f(t - iT_s) \qquad (3.13)$$

$$f(t) = \begin{cases} 1, & (-\Delta_G < t \le t_s) \\ 0, & (t \le -\Delta_G, \, t > t_s) \end{cases} \qquad (3.14)$$

$$f_k = \frac{(k-1)}{t_s}, \quad \Delta f = \frac{1}{t_s} \qquad (3.15)$$

where T_s, Δ_G and t_s are the OFDM symbol period, guard interval length, and observation period (often called "useful symbol length"), respectively, and they satisfy the following equation:

$$T_s = \Delta_G + t_s \qquad (3.16)$$

Although we will give the details in Chapter 8, we can see several applications of OFDM in real and commercial radio systems such as broadcasting and communications. The OFDM waveform of all the systems is mathematically expressed by (3.13) to (3.16) in the sense that it is transmitted and received with the IDFT and DFT and that it has a guard interval. Now, call it "the current form of OFDM." Figure 3.8 shows the cyclic extension technique, frequency spectrum of pulse waveform, and frequency spectrum of transmitted signal in the current form of OFDM.

For the OFDM signal, the total symbol transmission rate is given by

$$R = 1/T = N_{SC}/T_s \qquad (3.17)$$

and the bandwidth in terms of main lobe is written as

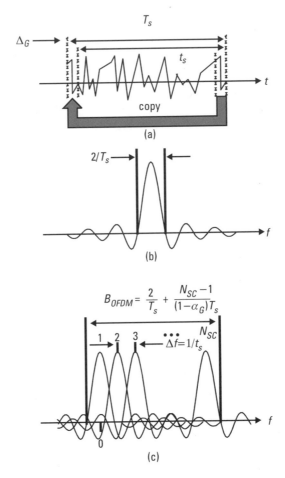

Figure 3.8 Current form of OFDM: (a) cyclic extension technique; (b) frequency spectrum of pulse waveform; and (c) frequency spectrum of OFDM signal.

$$B_{OFDM} = \frac{2}{T_s} + \frac{N_{SC} - 1}{(1 - \alpha_G) T_s} \qquad (3.18)$$

where α_G is the guard interval factor, which is defined as

$$\alpha_G = \frac{\Delta_G}{T_s} \qquad (3.19)$$

When the number of subcarriers is large, the bandwidth of the OFDM signal normalized by R is written as

$$\frac{B_{OFDM}}{R} = \frac{1}{1 - \alpha_G} \qquad (3.20)$$

Figure 3.9 shows the transmitted power spectra normalized by the symbol transmission rate. The left-hand column shows the power spectra for the OFDM signals without guard interval, whereas the right-hand column shows those for the OFDM signals with 20% guard interval (namely, the current form of OFDM). As the number of subcarrier increases, the side lobe level, namely, the outband radiation, can be reduced. In addition, the guard interval insertion introduces an expansion of the transmitted bandwidth [see (3.20)].

3.5 Conclusions

OFDM has long been studied and implemented to combat transmission channel impairments. Its applications have been extended from high frequency (HF) radio communication to telephone networks, digital audio broadcasting, and digital television terrestrial broadcasting. The advantage of OFDM, especially in the multipath propagation, interference, and fading environment, makes the technology a promising alternative in digital broadcasting and communications.

OFDM has already been accepted for digital broadcasting by European DAB, European DVB-T, and Japanese ISDB-T, and the new wireless local-area network standards such as IEEE 802.11a, HIPERLAN/2 and MMAC. Chapter 8 discusses these in more detail. Currently, OFDM and its related topics are of great interest to researchers in universities and research laboratories all over the world.

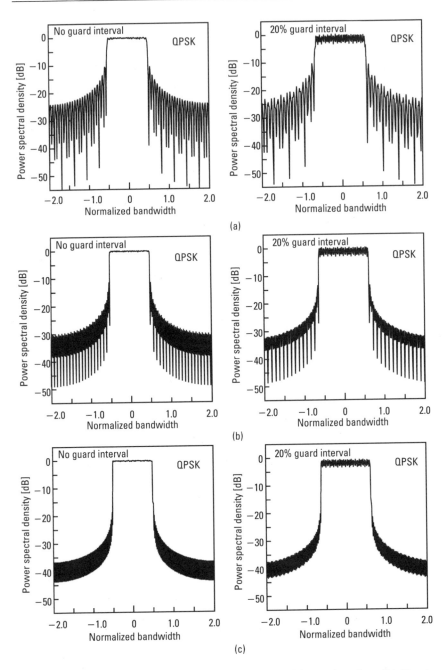

Figure 3.9 Transmitted power spectra of OFDM signals: (a) 16 subcarriers; (b) 64 subcarriers; and (c) 256 subcarriers.

References

[1] Prasad, R., *Universal Wireless Personal Communications*, Norwood, MA: Artech House, 1998.

[2] van Nee, R., and R. Prasad, *OFDM for Wireless Multimedia Communications*, Norwood, MA: Artech House, 2000.

[3] Mosier, R. R., and R. G. Clabaugh, "Kineplex, a Bandwidth-Efficient Binary Transmission System," *AIEE Trans.*, Vol. 76, January 1958, pp. 723–728.

[4] Zimmerman, M. S., and A. L. Kirsch, "The AN/GSC-10 (KATHRYN) Variable Rate Data Modem for HF Radio," *IEEE Trans. Commun. Technol.*, Vol. COM-15, April 1967, pp. 197–205.

[5] Marmuth, H. F., "On the Transmission of Information by Orthogonal Time Functions," *AIEE Trans; (communication and Electronics)*, Vol. 79, July 1960, pp. 248–255.

[6] Chang, R. W., "Synthesis of Band-Limited Orthogonal Signals for Multichannel Data Transmission," *Bell Sys. Tech. J.*, Vol. 45, December 1966, pp. 1775–1796.

[7] Saltzberg, B. R., "Performance of an Efficient Parallel Data Transmission System," *IEEE Trans. Commun.*, Vol. COM-15, No. 6, December 1967, pp. 805–813.

[8] Orthogonal Frequency Division Multiplexing," U.S. Patent No. 3,488,445, Filed Nov. 14, 1966, Issued Jan. 6, 1970.

[9] Weinstein, S. B., and P. M. Ebert, "Data Transmission by Frequency-Division Multiplexing Using the Discrete Fourier Transform," *IEEE Trans. Commun. Technol.*, Vol. COM-19, October 1971, pp. 628–634.

[10] Cimini, Jr., L. J., "Analysis and Simulation of a Digital Mobile Channel Using Orthogonal Frequency Division Multiplexing," *IEEE Trans. Commun.*, Vol. COM-33, No.7, July 1985, pp. 665–675.

[11] Alard, M., and R. Lassalle, "Principles of Modulation and Channel Coding for Digital Broadcasting for Mobile Receivers," *EBU Technical Review*, No. 224, 1989, pp. 168–190.

[12] Le Floch, B., R. Halbert-Lassalle, and D. Catelain, "Digital Sound Broadcasting for Mobile Receivers," *IEEE Trans. Consumer Electronics*, Vol. 73, 1989, pp. 30–34.

4

OFDM Characteristics

4.1 Introduction

The basic principle of OFDM, explained in Chapter 3, is to split a high-rate data stream into a number of lower rate streams that are transmitted simultaneously over a number of subcarriers. Because the symbol duration increases for the lower rate parallel subcarriers, the relative amount of dispersion in time caused by multipath delay spread is decreased, as introduced in Chapter 2. ISI is eliminated almost completely by introducing a guard interval in each OFDM symbol, where the OFDM signal is cyclically extended to avoid intersubcarrier interference. This whole process of generating an OFDM signal and the reasoning behind it are described in Chapter 3.

In OFDM system development, it is necessary to understand its performance characteristics, which are given qualitatively in [1]. This chapter presents the performance analysis of the OFDM system in great detail. After Section 4.2 shows the radio channel model for the bit error rate (BER) analysis, Section 4.3 first analyzes the BER in an additive white Gaussian noise (AWGN) channel, and then Sections 4.4 and 4.5 extend it for various cases, namely, phase shift keying/coherent detection (CPSK)- and PSK/differential detection (DPSK)-based OFDM systems in Rayleigh fading channels, respectively. Then, after introducing the robustness against frequency selective fading in Section 4.6, Section 4.7 discusses the robustness against man-made noises. Sections 4.8 and 4.9 analyze the sensitivity to frequency offset and nonlinear amplification, respectively. The sensitivity to analog to digital (A/D) and digital to analog (D/A) resolutions is investigated in Section 4.10. Section 4.11 concludes the chapter.

Before discussing the performance analysis of an OFDM system, this section gives the definition of an OFDM signal again. As derived in Chapter 3, the transmitted OFDM signal is defined as

$$s(t) = \sum_{i=-\infty}^{+\infty} \sum_{k=1}^{N_{SC}} c_{ki} e^{j2\pi f_k(t-iT_s)} f(t - iT_s) \tag{4.1}$$

$$f(t) = \begin{cases} 1, & (-\Delta_G < t \le t_s) \\ 0, & (t \le -\Delta_G, \, t > t_s) \end{cases} \tag{4.2}$$

$$f_k = \frac{(k-1)}{t_s}, \; \Delta f = \frac{1}{t_s} \tag{4.3}$$

$$T_s = \Delta_G + t_s \tag{4.4}$$

$$R = 1/T = N_{SC}/T_s \tag{4.5}$$

where N_{SC}, c_{ki}, f_k, T_s, Δ_G, t_s and $f(t)$ are the number of subcarriers, the ith information symbol at the kth subcarrier, the frequency of the kth subcarrier, the OFDM symbol period, the guard interval length, the observation period often called "useful symbol length," and the rectangular pulse waveform of the symbol, respectively. In addition, $R \, (= 1/T)$ is the total symbol transmission rate.

When we limit our interest only within binary PSK (BPSK) or quadrature PSK (QPSK) at all the subcarriers, the information symbol is given by

$$c_{ki} \in \left\{ e^{j\frac{2\pi m}{M}} \mid m = 0, 1, \ldots, M - 1 \right\} \tag{4.6}$$

where $M = 2^{k_M}$, so $k_M = 1$ for BPSK and $k_M = 2$ for QPSK.

4.2 Radio Channel Model

Chapter 2 described only the distortion that a transmitted signal experiences through a multipath fading channel. In a radio channel, generally, a transmitted signal is not only distorted by multipath fading but also corrupted by thermal noise, as shown in Figure 4.1. The received signal through the channel is written as

$$r(t) = \int_{-\infty}^{\infty} h(\tau; t) s(t - \tau) \, d\tau + n(t) \tag{4.7}$$

where $h(\tau; t)$ is the impulse response of the channel given by (2.4) and $n(t)$ is an AWGN with two-sided power spectral density of $N_0/2$.

Chapter 2 also showed that $h(\tau; t)$ is a random property; in other words, it is like a noise. Equation (4.7) shows that through a radio channel, a transmitted signal is multiplied by $h(\tau; t)$ and then is added by $n(t)$, therefore, we can consider that fading is a *multiplicative* noise.

If the channel impulse response is a time-invariant constant given by

$$h(\tau; t) = h\delta(\tau) \tag{4.8}$$

where h is a complex-valued channel gain, then we can ignore the effect of fading and there is only an AWGN in the channel. We call it "an AWGN channel." On the other hand, if the channel impulse response has a time-variant property, then there is fading and an AWGN in the channel. We call it "a fading channel."

The BER performance of a modulation/demodulation scheme largely depends on the received signal to noise (power) ratio (SNR) per bit, which is also called "ratio of energy per bit to power spectral density of noise (E_b/N_0)" or "the energy contrast." In the following sections, we will show the detailed BER performance of OFDM systems in the AWGN and fading channels.

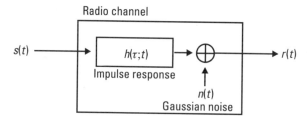

Figure 4.1 Radio channel model.

4.3 Bit Error Rate in AWGN Channel

Figure 4.2 shows a CPSK-based OFDM transmission system. Assuming that the receiver knows the DFT window timing perfectly, the DFT output at the nth subcarrier in $[iT_s, iT_s + t_s]$ is written as

$$r_{ni} = \frac{1}{t_s} \int_{iT_s}^{iT_s+t_s} r(t)\, e^{-j2\pi f_n(t-iT_s)}\, dt \tag{4.9}$$

Defining n_{ni} as the noise component in the DFT output, substituting (4.1), (4.7), and (4.8) into (4.9) leads to:

$$r_{ni} = h c_{ni} + n_{ni} \tag{4.10}$$

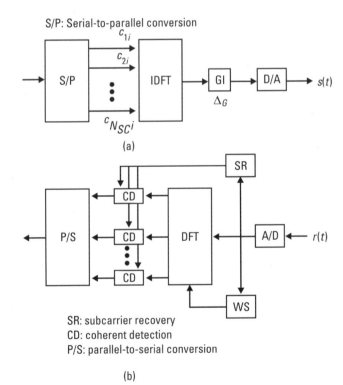

(a)

(b)

SR: subcarrier recovery
CD: coherent detection
P/S: parallel-to-serial conversion

Figure 4.2 Transmission diagram of a CPSK-based OFDM system: (a) transmitter; and (b) receiver.

Equation (4.9) clearly shows that the received signal is integrated only over the useful symbol period. For the current form of an OFDM system, the signal is transmitted even in the guard interval, so its power is not used for detection. Therefore, in the AWGN channel, the BER of a CPSK-based OFDM system is all the same as that of a CPSK-based SCM system [2], but we need to take into consideration "the power loss associated with guard interval insertion." When employing BPSK or QPSK at all the subcarriers, the BER is given by

$$P_{b,AWGN}^{B,\,coherent} = P_{b,AWGN}^{Q,\,coherent} = \frac{1}{2}\operatorname{erfc}\left(\sqrt{\gamma_b'}\,\right) \qquad (4.11)$$

where $\operatorname{erfc}(x)$ is the complementary error function given by

$$\operatorname{erfc}(x) = \frac{2}{\sqrt{\pi}} \int\limits_{x}^{\infty} e^{-t^2}\,dt \qquad (4.12)$$

γ_b' is the effective SNR per bit. Defining the SNR per bit as γ_b, it is written as

$$\gamma_b' = \frac{t_s}{T_s}\,\gamma_b = (1 - \alpha_G)\,\gamma_b \qquad (4.13)$$

Figure 4.3 shows a DPSK-based OFDM transmission system. The DPSK-based OFDM system is advantageous over the CPSK-based OFDM system, because differential detection can totally eliminate an elaborate subcarrier recovery that coherent detection requires, although the transmission performance of the DPSK-based system is inferior to that of the CPSK-based system.

The BER of a DPSK-based OFDM system is also the same as that of a DPSK-based SCM system [2]. When employing BPSK at all the subcarriers, the BER is given by

$$P_{b,AWGN}^{B,\,differential} = \frac{1}{2}e^{-\gamma_b'} \qquad (4.14)$$

and when employing QPSK, the BER is given by

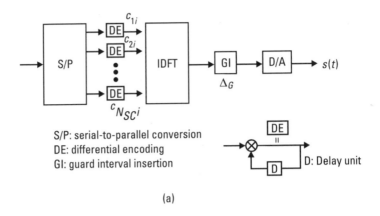

S/P: serial-to-parallel conversion
DE: differential encoding
GI: guard interval insertion

(a)

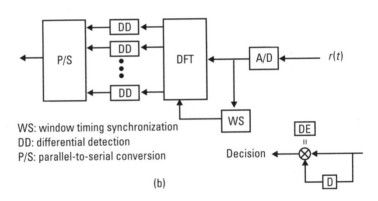

WS: window timing synchronization
DD: differential detection
P/S: parallel-to-serial conversion

(b)

Figure 4.3 Transmission diagram of a DPSK-based OFDM system: (a) transmitter; and (b) receiver.

$$P_{b,AWGN}^{Q, differential} = Q(a, b) - \frac{1}{2} I_0(ab) e^{-\frac{1}{2}(a^2 + b^2)} \tag{4.15}$$

$$a = \sqrt{\frac{\gamma_b'}{2}} \left(\sqrt{2 + \sqrt{2}} - \sqrt{2 - \sqrt{2}} \right) \tag{4.16}$$

$$b = \sqrt{\frac{\gamma_b'}{2}} \left(\sqrt{2 + \sqrt{2}} + \sqrt{2 - \sqrt{2}} \right) \tag{4.17}$$

In (4.15), $Q(a, b)$ is the Q-function given by

$$Q(a, b) = e^{-\frac{1}{2}(a^2 + b^2)} \sum_{j=0}^{\infty} \left(\frac{a}{b}\right)^j I_j(ab) \qquad (4.18)$$

where $I_j(x)$ is the jth order modified Bessel function of the first kind.

4.4 Bit Error Rate of CPSK-Based OFDM System in Rayleigh Fading Channels

In Rayleigh fading channels, the BER performance of a CPSK-based OFDM system depends on a subcarrier recovery method employed. We will discuss subcarrier recovery methods and their attainable BER performance in detail in Chapters 5 and 6, so here we will show a BER lower bound when assuming a perfect subcarrier recovery.

The p.d.f. of Rayleigh faded envelope is given by (2.16), so by substituting $\gamma'_b = \xi^2/2/\sigma_n^2$ and $\overline{\gamma'_b} = \sigma_r^2/\sigma_n^2$ (average E_b/N_0) into (2.16), we can rewrite the p.d.f. of γ'_b as (σ_n^2 is the noise power)

$$p(\gamma'_b) = \frac{1}{\overline{\gamma'_b}} e^{-\frac{\gamma'_b}{\overline{\gamma'_b}}} \qquad (4.19)$$

Therefore, by averaging (4.11) with (4.19), we can obtain the BER in the Rayleigh fading channel as

$$P_{b, fading}^{B, coherent} = P_{b, fading}^{Q, coherent} = \int_0^{\infty} \frac{1}{2} \mathrm{erfc}\left(\sqrt{\gamma'_b}\right) p(\gamma'_b) \, d\gamma'_b \qquad (4.20)$$

$$= \frac{1}{2}\left(1 - \sqrt{\frac{\overline{\gamma'_b}}{1 + \overline{\gamma'_b}}}\right)$$

This equation is valid as the lower bound regardless of the frequency selectivity and time selectivity of the channel.

4.5 Bit Error Rate of DPSK-Based OFDM System in Rayleigh Fading Channels

Similarly as in the derivation in Section 4.4, we could obtain the BER for DPSK in Rayleigh fading channels, but we are actually successful only for the BER lower bound of a BDPSK-based OFDM system as

$$P_{b,\,fading}^{B,\,differential} = \int_0^\infty \frac{1}{2} e^{-\gamma_b'} \, p(\gamma_b') \, d\gamma_b' \qquad (4.21)$$

$$= \frac{1}{2}\left(1 - \frac{\overline{\gamma_b'}}{1 + \overline{\gamma_b'}}\right)$$

so we need a more detailed discussion for the BER derivation.

4.5.1 Theoretical Bit Error Rate Analysis

In differentially encoded PSK, phase transition between successive symbols conveys information. Defining the phase difference between the ith and $(i - 1)$th symbols as $\Delta\phi_{ki}$, for BDPSK, it is given by

$$\Delta\phi_{ki} = \begin{cases} 0, & (b_{ki} = 0) \\ \pi, & (b_{ki} = 1) \end{cases} \qquad (4.22)$$

on the other hand, for QDPSK,

$$\Delta\phi_{ki} = \begin{cases} 0, & (b_{k(2i-1)}b_{k2i} = 00) \\ \dfrac{\pi}{2}, & (b_{k(2i-1)}b_{k2i} = 01) \\ \pi, & (b_{k(2i-1)}b_{k2i} = 10) \\ \dfrac{3\pi}{2}, & (b_{k(2i-1)}b_{k2i} = 11) \end{cases} \qquad (4.23)$$

where b_{ki} is the ith information symbol at the kth subcarrier. Figure 4.4 shows the phase transitions.

Assume that $b_{ni} = 0$ was transmitted at the nth subcarrier for BDPSK. A bit error occurs when the phase difference between the ith and $(i - 1)$th symbols at the nth DFT output is less than $-\pi/2$ and greater than $+\pi/2$, so the BER is given by

$$P_{b,\,fading}^{B,\,differential} = \Pr\left\{ \mathrm{Re}\left[\frac{r_{ni}\, r_{n(i-1)}^*}{|r_{ni}||r_{n(i-1)}|} \right] \le 0 \right\} \qquad (4.24)$$

On the other hand, assume that $b_{ni} = 00$ was transmitted for QDPSK. Similar to the case of BDPSK, a bit error occurs when the phase difference is less than $-\pi/4$ and greater than $+\pi/4$, so the BER is given by

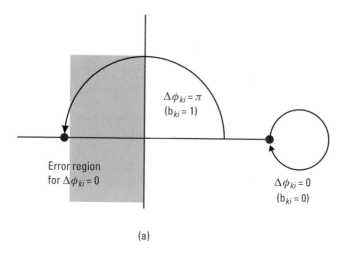

(a)

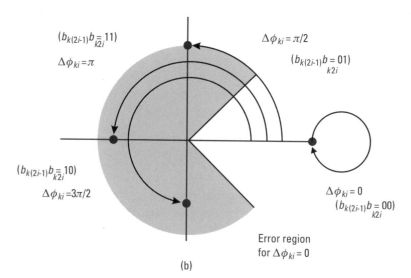

(b)

Figure 4.4 Phase transitions in DPSK signal constellations: (a) BDPSK; and (b) QDPSK.

$$P_{b, fading}^{Q, differential} = \frac{1}{k_M} \Pr \left\{ \mathrm{Re} \left[\frac{r_{ni} r_{n(i-1)}^*}{|r_{ni}||r_{n(i-1)}|} e^{j\frac{\pi}{4}} \right] \leq 0 \right. \tag{4.25}$$

$$\text{and } \mathrm{Re} \left[\frac{r_{ni} r_{n(i-1)}^*}{|r_{ni}||r_{n(i-1)}|} e^{-j\frac{\pi}{4}} \right] \leq 0 \left. \right\}$$

In (4.24) and (4.25), the denominators are always positive, so we have

$$P_{b,\,fading}^{B\,and\,Q,\,differential} = \Pr\{D_{ni} \leq 0\} \qquad (4.26)$$

$$D_{ni} = \mathrm{Re}\left[r_{ni} r_{n(i-1)}^* \, e^{-j\left(\frac{\pi}{2} - \frac{\pi}{M}\right)} \right] \qquad (4.27)$$

We can derive the BER as (see Appendix 4A) [3–11]

$$P_{b,\,fading}^{B\,and\,Q,\,differential} = \frac{1}{2}\left\{ 1 - \frac{\rho \sin\left(\dfrac{\pi}{M}\right)}{\sqrt{1 - \rho^2 \cos^2\left(\dfrac{\pi}{M}\right)}} \right\} \qquad (4.28)$$

where ρ is the magnitude of the normalized correlation between r_{ni} and $r_{n(i-1)}$, which is given by

$$\rho = \frac{\mathrm{Re}\,[E[r_{ni} r_{n(i-1)}^*\,]]}{E[r_{ni} r_{ni}^*]} \qquad (4.29)$$

We can see that the normalized correlation ρ plays a key role in the BER performance.

4.5.2 Bit Error Rate in Frequency Selective and Time Selective Rayleigh Fading Channels

As shown in Chapter 2, the impulse response of frequency selective fast Rayleigh fading channel can be modeled as a tapped delay line where each tap is an independent zero-average complex-valued Gaussian random process as

$$h(\tau;\,t) = \sum_{l=1}^{L_1+L_2} \alpha_l(t)\,\delta(\tau - \tau_l) \qquad (4.30)$$

where $\alpha_l(t)$ is the loss of the lth path, which is assumed as a complex-valued Gaussian random variable with average of zero and variance of σ_l^2, and τ_l is the propagation delay for the lth path, which is classified as follows:

$$0 < \tau_l \leq \Delta_G,\ (l = 1, \ldots, L_1) \qquad (4.31)$$

$$\Delta_G < \tau_l \le T_s, \, (l = L_1 + 1, \ldots, L_1 + L_2) \tag{4.32}$$

Substituting (4.1), (4.7) and (4.30) to (4.32) into (4.8) leads to:

$$r_{ni} =$$

$$\left\{ \sum_{l=1}^{L_1} \frac{c_{ni}}{t_s} e^{-j2\pi f_n \tau_l} \int_{iT_s}^{iT_s+t_s} \alpha_l(t) \, dt + \sum_{l=L_1+1}^{L_1+L_2} \frac{c_{ni}}{t_s} e^{-j2\pi f_n \tau_l} \int_{iT_s-\Delta_G+\tau_l}^{iT_s+t_s} \alpha_l(t) \, dt \right\}$$

$$+ \left\{ \sum_{l=1}^{L_1} \sum_{\substack{k=1 \\ k \ne n}}^{N_{SC}} \frac{c_{ki}}{t_s} e^{-j2\pi f_k \tau_l} \int_{iT_s}^{iT_s+t_s} \alpha_l(t) e^{-j2\pi(f_k-f_n)(t-iT_s)} \, dt \right.$$

$$+ \sum_{l=L_2+1}^{L_1+L_2} \sum_{\substack{k=1 \\ k \ne n}}^{N_{SC}} \frac{c_{ki}}{t_s} e^{-jf_k \tau_l} \int_{iT_s-\Delta_G+\tau_l}^{iT_s+t_s} \alpha_l(t) e^{-j2\pi(f_k-f_n)(t-iT_s)} \, dt$$

$$+ \left. \sum_{l=L_1+1}^{L_1+L_2} \sum_{\substack{k=1 \\ k \ne n}}^{N_{SC}} \frac{c_{k(i-1)}}{t_s} e^{-j2\pi f_k(\tau_l-T_s)} \int_{iT_s}^{iT_s-\Delta_G+\tau_l} \alpha_l(t) e^{-j2\pi(f_k-f_n)(t-iT_s)} \, dt \right\}$$

$$+ \sum_{l=L_1+1}^{L_1+L_2} \frac{c_{n(i-1)}}{t_s} e^{-j2\pi f_n(\tau_l-T_s)} \int_{iT_s}^{iT_s-\Delta_G+\tau_l} \alpha_l(t) \, dt + n_{ni}$$

$$\tag{4.33}$$

In (4.33), the first term means the desired signal component, the second and third terms mean the intersubcarrier interference and ISI, respectively, and the fourth term means the Gaussian noise component with average of zero and power of σ_n^2. σ_l^2 and σ_n^2 satisfies the following equation:

$$\overline{\gamma_b} = \frac{\displaystyle\sum_{l=1}^{L_1+L_2} \sigma_l^2}{k_M \sigma_n^2} \tag{4.34}$$

Substituting (4.33) into (4.29), we could derive the BER, but it would be untractable. Here, focusing our attention on a case where the receiver uses an omnidirectional antenna, ρ can be calculated in a closed form as (see Appendix 4B)

$$\rho = \frac{\sigma_{S1}^2}{\sigma_{S2}^2 + \sigma_I^2 + \sigma_n^2} \tag{4.35}$$

where σ_{S1}^2, σ_{S2}^2 and σ_I^2 are given by

$$\sigma_{S1}^2 = \sum_{l=1}^{L_1} \sigma_l^2 \left\{ 1 - (\pi f_D)^2 \left(\frac{t_s^2}{6} + T_s^2 \right) \right\} \tag{4.36}$$
$$+ \sum_{l=L_1+1}^{L_1+L_2} \sigma_l^2 \frac{(T_s - \tau_l)^2}{t_s^2} \left\{ 1 - (\pi f_D)^2 \left(\frac{(T_s - \tau_l)^2}{6} + T_s^2 \right) \right\}$$

$$\sigma_{S2}^2 = \sum_{l=1}^{L_1} \sigma_l^2 \left\{ 1 - (\pi f_D)^2 \frac{t_s^2}{6} \right\}$$
$$+ \sum_{l=L_1+1}^{L_1+L_2} \sigma_l^2 \frac{(T_s - \tau_l)^2}{t_s^2} \left\{ 1 - (\pi f_D)^2 \frac{(T_s - \tau_l)^2}{6} \right\} \tag{4.37}$$
$$+ \sum_{l=L_1+1}^{L_1+L_2} \sigma_l^2 \frac{(-\Delta_G + \tau_l)^2}{t_s^2} \left\{ 1 - (\pi f_D)^2 \frac{(-\Delta_G + \tau_l)^2}{6} \right\}$$

$$\sigma_I^2 = \sum_{l=1}^{L_1} \sum_{\substack{k=1 \\ k \neq n}}^{N_{SC}} \sigma_l^2 \frac{f_D^2 t_s^2}{2(k-n)}$$
$$+ \sum_{l=1}^{L_1} \sum_{\substack{k=1 \\ k \neq n}}^{N_{SC}} \sigma_l^2 \left[\frac{f_D^2 (T_s - \tau_l)^2}{2(k-n)} \cos\left(\frac{2\pi(k-n)(T_s - \tau_l)}{t_s} \right) \right.$$
$$+ \frac{f_D^2 t_s (T_s - \tau_l)^2}{\pi(k-n)} \sin\left(\frac{2\pi(k-n)(T_s - \tau_l)}{t_s} \right)$$
$$\left. + \left\{ \frac{1}{2\pi^2(k-n)^2} + \frac{3f_D^2 t_s^2}{4\pi^2(k-n)^4} \right\} \right.$$

$$\times \left\{ 1 - \cos\left(\frac{2\pi(k-n)(T_s - \tau_l)}{t_s} \right) \right\}$$

$$+ \frac{f_D^2(-\Delta_G + \tau_l)^2}{2(k-n)^2} \cos\left(\frac{2\pi(k-n)(-\Delta_G + \tau_l)}{t_s} \right)$$

$$+ \frac{f_D^2 t_s(-\Delta_G + \tau_l)^2}{\pi(k-n)} \sin\left(\frac{2\pi(k-n)(-\Delta_G + \tau_l)}{t_s} \right)$$

$$+ \left\{ \frac{1}{2\pi^2(k-n)^2} + \frac{3f_D^2 t_s^2}{4\pi^2(k-n)^4} \right\}$$

$$\times \left\{ 1 - \cos\left(\frac{2\pi(k-n)(-\Delta_G + \tau_l)}{t_s} \right) \right\} \Bigg] \Bigg]$$

$$(4.38)$$

In (4.36) to (4.38), σ_{S1}^2 is the power of the signal, σ_{S2}^2 is the power of the ISI and σ_I^2 is the power of the intersubcarrier interference. Note that σ_I^2 is a function of n. This means that the power of intersubcarrier interference depends on the position of subcarrier that we pay attention to. Therefore, to calculate the BER in a strict sense, we need to average (4.28) with n.

Equations (4.35) to (4.38) are valid for frequency selective and time selective Rayleigh fading channels. Therefore, when setting f_D to be zero, we can obtain the BER in a frequency selective slow (time nonselective) Rayleigh fading channel, whereas when setting L_2 to be one, we can obtain the BER in a frequency nonselective (flat) time selective (fast) Rayleigh fading channel. To have the BER in a frequency nonselective slow Rayleigh fading channel as the lower bound, setting $f_D \to 0$ and $L_2 \to 1$ in (4.35) to (4.38) leads to:

$$\rho = \frac{1}{1 + \sigma_n^2 / ((1 - \alpha_G)\sigma_1^2)} = \frac{k_M \overline{\gamma_b'}}{k_M \overline{\gamma_b'} + 1} \qquad (4.39)$$

On the other hand, when $\rho \cong 1.0$, we can approximate (4.28) as

$$P_{b, fading}^{B \, and \, Q, \, differential} = \frac{1}{2}(1 - \rho) \qquad (4.40)$$

so substituting (4.39) into (4.40) leads to:

$$P_{b, fading}^{B \, and \, Q, \, differential} = \frac{1}{2} \left(1 - \frac{k_M \overline{\gamma_b'}}{k_M \overline{\gamma_b'} + 1} \right) \tag{4.41}$$

This equation is equivalent to (4.22) for BDPSK.

4.5.3 Optimum Number of Subcarriers and Optimum Length of Guard Interval

When the symbol transmission rate, channel frequency selectivity, and channel time selectivity are given, the transmission performance becomes more sensitive to the time selectivity as the number of subcarriers increases because the wider symbol duration is less robust to the random FM noise, whereas it becomes poor as the number of subcarriers decreases because the wider power spectrum of each subcarrier is less robust to the frequency selectivity. Figure 4.5 shows the relation among the number of subcarriers, frequency selectivity, and time selectivity.

On the other hand, the transmission performance becomes poor as the length of guard interval increases because the signal transmission in the guard interval introduces a power loss, whereas it becomes more sensitive to the frequency selectivity as the length of guard interval decreases because the shorter guard interval is less robust to the delay spread. Figure 4.6 shows the relation among the length of guard interval, energy efficiency, and frequency selectivity.

Therefore, for the given R, τ_{RMS} and f_D, there exist the optimum values in N_{SC} and Δ_G, and we can determine it by maximizing the correlation given by (4.35) to (4.38), because it means a measure to show how much distortion the frequency and time selective fading channel gives to the transmitted signal [10]:

$$[N_{SC}, \Delta_G]_{opt} = \arg \left\{ \max \, \rho \left(N_{SC}, \Delta_G \, | \, R, \tau_{RMS}, f_D \right) \right\} \tag{4.42}$$

Figure 4.7 shows the optimum number of subcarriers and optimum length of guard interval as a function of the maximum Doppler frequency [see Figure 4.7(a)] and as a function of the RMS delay spread [see Figure 4.7(b)]. Here, we assume an exponentially decaying multipath delay profile with 20 paths and normalize the Doppler frequency with the total symbol transmission rate R whereas the RMS delay spread with the total symbol transmission period T.

Figure 4.8 shows the attainable BER of a QDPSK-based OFDM system in Rayleigh fading channels, where we use the optimized N_{SC} and Δ_G and

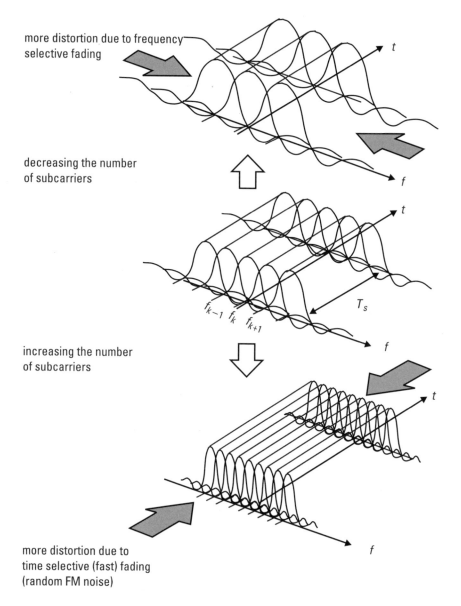

more distortion due to frequency selective fading

decreasing the number of subcarriers

increasing the number of subcarriers

more distortion due to time selective (fast) fading (random FM noise)

Figure 4.5 Relation among number of subcarriers, frequency selectivity, and time selectivity.

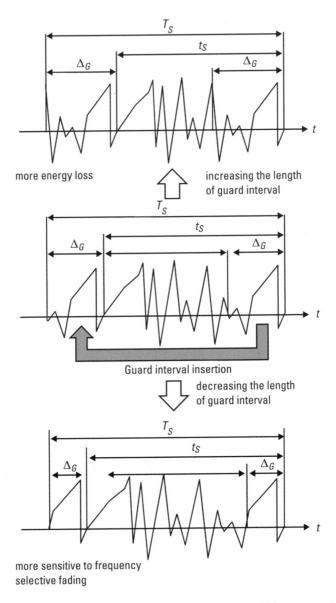

Figure 4.6 Relation among length of guard interval, energy efficiency, and frequency selectivity.

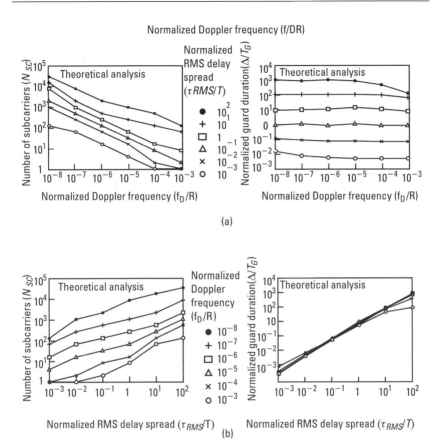

Figure 4.7 Optimum number of subcarriers and optimum length of guard interval: (a) optimum number of subcarriers and optimum length of guard interval against Doppler frequency; and (b) optimum number of subcarriers and optimum length of guard interval against RMS delay spread.

set the average E_b/N_0 to be 40 dB. When $[\tau_{RMS}/T, f_D/R]$ of the channel lies in the hatched region (roughly, $\tau_{RMS} \cdot f_D \leq 1.0 \times 10^{-6}$), we can make the BER be less than $10^{-4.25} = 5.62 \times 10^{-5}$. Note that the BER of a QDPSK SCM system in a frequency nonselective slow Rayleigh fading channel is $10^{-4.30} = 5.01 \times 10^{-5}$ at the average E_b/N_0 of 40 dB (the BER lower bound). This means that the BER degradation is less than 12.2%.

4.5.4 Numerical Results and Discussions

Table 4.1 shows the transmission parameters to demonstrate the performance of OFDM systems.

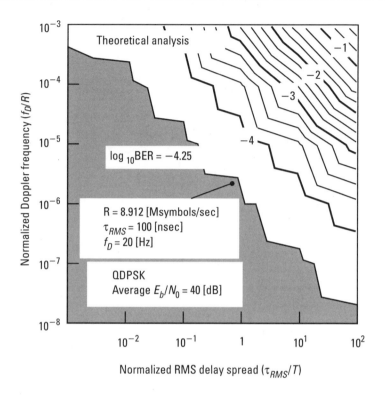

Figure 4.8 Attainable BER by multicarrierization.

Table 4.1
Transmission Parameters for BER Evaluation in Rayleigh Fading Channels

Total symbol transmission rate (R)	8.192 [Msymbols/sec]
Multipath delay profile ($\phi_H(\tau)$)	Exponentially decaying 20 paths
RMS delay spread (τ_{RMS})	100 [nsec]
Maximum Doppler frequency (f_D)	20 [Hz]

Figure 4.9 shows the BER of a QDPSK-based OFDM system versus the average E_b/N_0 in a frequency selective fast Rayleigh fading channel, where $N_{SC} = 128$ and $\Delta_G/T_s = 0.06$ are optimized for the transmission parameters shown in Table 4.1. In this figure, the BER of a QDPSK-based SCM system in a frequency nonselective slow fading channel is also shown as the BER lower bound, which is given by (4.40). The theoretical analysis and computer simulation results are in complete agreement, and there is little difference in the BER between the optimized OFDM system and the lower bound.

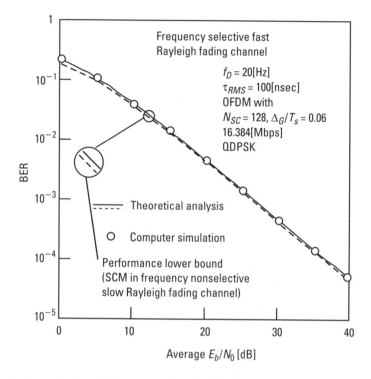

Figure 4.9 BER of a QDPSK-based OFDM system in a frequency selective fast Rayleigh fading channel.

Figure 4.10 shows the BER of the optimized QDPSK-based OFDM system in a frequency nonselective slow Rayleigh fading channel. There is also little difference in the BER between the optimized OFDM system and the lower bound. This means that the OFDM system optimized for the channel maximum Doppler frequency/RMS delay spread can keep the minimum BER even when the channel time selectivity and frequency selectivity disappear.

Figure 4.11 shows the BER of a BDPSK-based OFDM system optimized for the transmission parameters shown in Table 4.1. There is also little difference in the BER between the optimized OFDM system and the lower bound.

4.6 Robustness Against Frequency Selective Fading

In Section 4.5, we saw the BER performance of DPSK-based OFDM systems in Rayleigh fading channels. Here, we will discuss the robustness of an OFDM system against frequency selective fading.

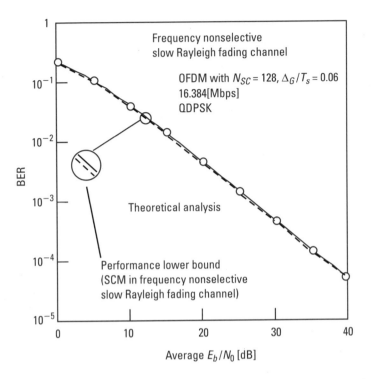

Figure 4.10 BER of the optimized QDPSK-based OFDM system in a frequency nonselective slow Rayleigh fading channel.

Assume a time-invariant impulse response of a channel that has a support shorter than the guard interval (in other words, the largest time delay is smaller than the length of the guard interval). In this case, the impulse response is written as

$$
h(\tau; t) = \begin{cases} h(\tau), & (0 < \tau \le \Delta_G) \\ 0, & (\tau \le 0, \ \tau > \Delta_G) \end{cases}
\tag{4.43}
$$

Substituting (4.1), (4.7) and (4.43) into (4.8), we have

$$
r_{ni} = H_n c_{ni} + n_n
\tag{4.44}
$$

$$
H_n = \int_0^{\Delta_G} h(\tau) e^{-j2\pi f_n \tau} \, d\tau
\tag{4.45}
$$

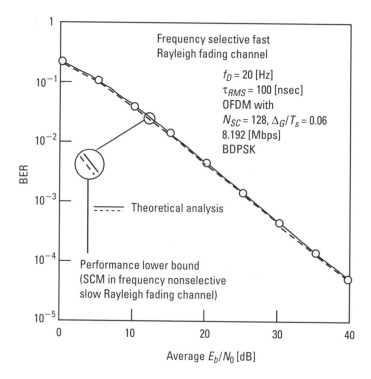

Figure 4.11 BER of a BDPSK-based OFDM system in a frequency selective fast Rayleigh fading channel.

Equation (4.44) clearly shows that if the guard interval absorbs all the time delays of the channel, no ISI and no intersubcarrier interference due to multipath propagation occurs. It is also interesting to compare (4.44) with (3.4).

4.7 Robustness Against Man-Made Noises

In recent years, it has been reported that a variety of man-made noises, which are mostly non-Gaussian, namely, impulsive in nature, have drastically deteriorated the performance of digital communications systems, such as asymmetrical digital subscriber lines (ADSL) [12], digital broadcasting [13], and 2.4-GHz wireless LANs [14]. These communications systems have already adopted OFDM as their physical layer standards, so it must be important to discuss the performance of OFDM systems in impulsive man-made noises.

An OFDM system replaces the use of a bank of bandpass filters by that of the DFT, so it could give us an impression that it is robust to impulsive noises, because it can split the spectrally widespread noise energy into many subbands. Indeed, the robustness of an OFDM system against impulsive noises has been long believed [15–18]. However, the effect of impulsive noise in SCM systems has been intensively investigated so far [19, 20], whereas its effect on OFDM systems has not yet been well studied [21].

In this section, we show the BER performance of an OFDM system in a "generalized shot noise (GSN)" channel [22, 23]. Here, to discuss the effect of GSN on the BER, we do not take channel fading into consideration. Furthermore, we assume a BCPSK format and perfect DFT window synchronization and subcarrier recovery.

4.7.1 Generalized Shot Noise Channel

Figure 4.12 shows the GSN, which is often called "impulsive noise." The complex-valued GSN is written as (in equivalent baseband expression) [20]

$$n_{GS}(t) = \sum_{m=-\infty}^{\infty} \Gamma_m e^{j\phi_m} \delta(t - t_m) \qquad (4.46)$$

where t_m is the arrival time of impulses that belong to a set of Poisson points with average number of impulses per second λ_{GS}, Γ_m is an i.i.d. random amplitude with p.d.f. of $p(\Gamma_m)$, and ϕ_m is an i.i.d. random phase with uniform p.d.f. over $[0, 2\pi]$. Assume that Γ_m and ϕ_m are statistically independent. The power spectrum of the GSN should be white with spectral density:

$$N_{GS} = \frac{1}{2} E[\Gamma_m^2] \lambda_{GS} \qquad (4.47)$$

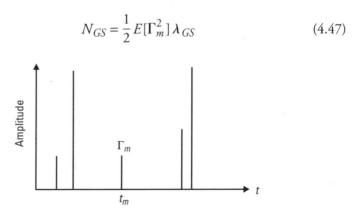

Figure 4.12 GSN.

A lot of measurements have revealed that $p(\Gamma_m)$ is one of various forms of *heavy-tailed* distribution functions, such as lognormal [13, 20] and $1/x^v$ [24, 25]. Heavy-tailed distribution functions decay slowly and their impulsive characteristics are more interesting than those of Gaussian with large variance and Rayleigh distributions, which have been commonly used in most analysis papers.

4.7.2 Bit Error Rate of SCM in GSN Channel

For SCM system, the transmitted signal is written as (in the equivalent baseband expression)

$$s(t) = \sum_{i=-\infty}^{\infty} c_i \psi(t - iT) \tag{4.48}$$

where c_i and $\psi(t)$ denote the ith data symbol (+1 or −1) and the basis function used as the transmitting filter, respectively. The transmitting filter adopted here is the well-known root-Nyquist filter with roll-off factor of α_{roll}, and its energy is normalized to be unity.

Assume a case when the transmitted signal is disturbed by the AWGN and GSN. The received signal is then written as

$$r(t) = \sqrt{2E_b}\, s(t) + n_G(t) + n_{GS}(t) \tag{4.49}$$

where E_b is the signal energy per bit and $n_G(t)$ is the complex-valued Gaussian noise. The received signal is then applied to the matched filter (root-Nyquist filter) to recover the desired data symbols.

Figure 4.13 shows how to evaluate the BER. Without loss of generality, we will pay attention to the probability of bit error at sampling instant $t = 0$. From the figure, it can be seen that the effect of GSN depends on where the impulses are present, namely, t_m. We heuristically assume that the impulses that occur outside the interval W_o do not introduce bit errors, so we can only be concerned with the impulses that occur within the interval W_o. For example, around 99.8% of the total energy of root-Nyquist filtered pulse with $\alpha_{roll} = 0.25$ falls in the interval $\pm 3T$ from the center of the pulse, therefore, we can set $W_o = 6T$. In addition, we assume that t_m is uniformly distributed over W_o [26].

The inphase component of the matched filter output at sampling time $t = 0$ is written as

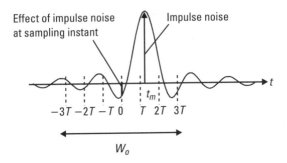

Figure 4.13 Effect of impulsive noise on sampling instant.

$$y = \text{Re}\left[\int_{-\infty}^{\infty} r(t)\,\psi(t)\,dt\right] = \sqrt{2E_b}\,c_0 + n'_G + n'_{GS} \qquad (4.50)$$

where the AWGN component n'_G is the Gaussian random variable with an average of zero and power of N_0 and the GSN component n'_{GS} is written as

$$n'_{GS} = \sum_m \Gamma_m \cos\phi_m\,\psi(t_m) \qquad (4.51)$$

where the summation with index m is done for all impulses that occur in the interval W_o.

Define β as the ratio of signal energy per bit to background Gaussian noise power spectral density, E_b/N_0, and normalize the y in (4.49) by $\sqrt{2E_b}$. Then, the normalized receiver output is written as

$$y' = c_0 + \frac{n'_G}{\sqrt{2E_b}} + \frac{1}{\sqrt{\beta}}\sum_m \gamma_m \cos\phi_m\,\psi'(t_m) \qquad (4.52)$$

where the newly defined variables γ_m and $y'(t)$ denote $\Gamma_m/\sqrt{2N_0 T}$ and $\sqrt{T}\psi(t_m)$, respectively, and γ_m is still an i.i.d. random variable with p.d.f. similar to Γ_m. Note that the normalization can avoid dependency on the symbol duration in the obtained results.

Now, we define p_j and $P_b(j)$ as the probability that exactly j impulses occur in the observation interval W_o and the conditional bit error probability given that j impulses occur in the observation interval W_o. The BER is then written as

$$P_b = \sum_{j=0}^{\infty} p_j P_b(j) \cong p_0 P_b(0) + (1 - p_0) P_b(1) \tag{4.53}$$

where $P_b(0)$ is the probability of bit error when no impulse occurs in W_o and is equal to $0.5 \, \mathrm{erfc}(\sqrt{\beta})$. Generally, an SCM system satisfies the condition of $\lambda_{GS} T$ (also $\lambda_{GS} W_o$) $\ll 1$, so the BER is well approximated by (4.52). The same BER is obtained for $c_0 = +1$ or -1, so we consider only the case for $c_0 = +1$. Finally, p_j and $P_b(1)$ are given by

$$p_j = \frac{(\lambda_{GS} W_o)^j e^{-\lambda_{GS} W_o}}{j!} \tag{4.54}$$

$$P_b(1) = \frac{1}{2\pi W_o} \int_{-W_o/2}^{W_o/2} \int_0^{2\pi} \int_0^{\infty} 0.5 \, \mathrm{erfc}\left\{ \left(1 + \frac{1}{\sqrt{\beta}} \, \psi'(t_1) \, \gamma_1 \cos \phi_1 \right) \sqrt{\beta} \right\}$$

$$= p(\gamma_1) \, d\gamma_1 \, d\phi_1 \, dt_1$$

$$\tag{4.55}$$

4.7.3 Bit Error Rate of OFDM in GSN Channel

The OFDM signal is given by (4.1) to (4.5) and the DFT output at the nth subcarrier in $[iT_s, iT_s + t_s]$ is given by (4.10). Without loss of generality, we can drop the subscript i, then the inphase component of the nth subcarrier is written as

$$y_n = \mathrm{Re}\left[\int_0^{t_s} \{ \sqrt{2E_b} \, s(t) + n_G(t) + n_{GS}(t) \} e^{-j2\pi f_n t} \, dt \right] \tag{4.56}$$

$$= (1 - \alpha_G) \sqrt{2E_b T_b} \, c_n + n_{Gn} + n_{GSn}$$

where n_{Gn} is the AWGN component at the nth subcarrier, which is a Gaussian random variable with an average of zero and power of $N_0 t_s$, whereas n_{GSn} is the GSN component of the nth subcarrier, which is written as

$$
\begin{aligned}
n_{GSn} &= \mathrm{Re}\left[\int_0^{t_s} \sum_m \Gamma_m e^{j\phi_m}\, \delta(t - t_m)\, e^{-j2\pi f_n t}\, dt\right] \\
&= \sum_m \Gamma_m \cos(2\pi f_n t_m - \phi_m) \qquad\qquad (4.57) \\
&= \sum_m \Gamma_m \cos\theta_m
\end{aligned}
$$

In (4.57), we have taken into consideration the fact that the phase in $\cos(*)$ can be replaced by a uniformly distributed random variable over $[0, t_s]$, namely, θ_m.

The analysis from now is similar to that of an SCM system, so we will be brief. We normalize y_n in (4.56) by $(1 - \alpha_G)\sqrt{2E_b T_b}$, so that the normalized nth DFT output is written as

$$
\begin{aligned}
y_n' &= c_n + n_{Gn}' + n_{GSn}' \qquad\qquad (4.58) \\
&= c_n + n_{Gn}' + \frac{\kappa}{\sqrt{\beta}} \sum_m \gamma_m \cos\theta_m
\end{aligned}
$$

where the newly defined variable κ is $1/\big((1 - \alpha_G)\sqrt{N_{SC}}\big)$.

The BER can be written as (4.53). However, we cannot neglect larger j, as done in the analysis of the SCM system, because the average number of impulses per OFDM symbol $(\lambda_{GS} t_s)$ linearly increases as the number of subcarriers (N_{SC}) increases. Therefore, p_j and $P_b(j)$ are finally written as

$$
p_j = \frac{(\lambda_{GS} t_s)^j e^{-\lambda_{GS} t_s}}{j!} \qquad\qquad (4.59)
$$

$$
P_b(j) = \frac{1}{(2\pi)^j} \int_0^\infty\int_0^{2\pi} \cdots \int_0^\infty\int_0^{2\pi}
$$

$$
0.5\,\mathrm{erfc}\left\{\left(1 + \frac{\kappa}{\sqrt{\beta}}(\gamma_1\cos\phi_1 + \ldots + \gamma_j\cos\phi_j)\sqrt{(1 - \alpha_G)\beta}\right\}
$$

$$
\times p(\gamma_1)\ldots p(\gamma_j)\, d\gamma_1\, d\theta_1 \ldots d\gamma_j\, d\theta_j \qquad\qquad (4.60)
$$

The numerical computation of $P_b(j)$ with the above direct integration is time consuming and its result converges slowly; furthermore, it would be very difficult to calculate it accurately when j is more than 2. However, fortunately, when j becomes larger, the p.d.f. of the sum of the AWGN and GSN components tends to be Gaussian. With the Gaussian approximation, (4.60) will be reduced to a simple closed form; namely, it contains only one term of erfc(*).

As shown, the direct evaluation of (4.60) is somehow tough work. To exclude the integration problem, we introduce a new elegant method called "saddle-point method" for calculating $P_b(j)$. The discussion here is brief; one can find more details on this method in [27].

Let $h(s) = E[e^{-sy_n'}]$ denote the moment generating function (m.g.f.) of y_n' given by (4.58). Taking into consideration all the components in (4.58) are statistically independent and assuming $c_n = 1$, then $P_b(j)$ can be written as

$$P_b(j) = \frac{1}{2\pi j} \int_{s_0-j\infty}^{s_0+j\infty} s^{-1} h(s) \, ds = \frac{1}{\pi} \int_0^\infty \mathrm{Re}\left[s^{-1} h(s)\big|_{s=s_0+jy} \right] dy$$

(4.61)

$$h(s) = e^{-s} e^{\frac{s^2}{4(1-\alpha_G)\beta}} g(s)^j$$

(4.62)

where $g(s)$ denotes the m.g.f. of $\left(\kappa / \sqrt{\beta} \right) \gamma_m \cos \theta_m$ [see (4.58)]. We can evaluate this integration by the numerical quadrature passing straightly through the positive real-valued saddle-point s_0 (also see how to find s_0 in [27]). Even though this path is not an optimum one, namely, not an exactly steepest descent path, it still gives a very fast convergence. There appears only one integration in (4.60), which makes the calculation of $P_b(j)$ easy and effective for every j. On the other hand, this method has one drawback, that is, $g(s)$ must be known in a closed form.

4.7.4 Numerical Results and Discussions

The effect of impulsive noise on symbol decision depends on the amplitude distribution of real impulsive noise and the receiving filter, and, in practice, the heavy-tailed distribution of the amplitude varies from system to system and might not be fitted perfectly into well-known functions. However, to understand the robustness of an OFDM system against impulsive noises, it is enough to show two examples for impulsive noises; a lognormal distribution of γ_m as a measured example and a Laplace distribution of $\eta_m = \gamma_m \cos \theta_m$ as a simplest mathematical example [23]. We can apply the saddle-point method for the Laplacian example. In the following numerical results, we assume $\alpha_{roll} = 0.25$ and $W_o = 6T$. Furthermore, we set $\alpha_G = 0$ to eliminate energy inefficiency of an OFDM system. Table 4.2 shows the transmission parameters to demonstrate the performance of OFDM and SCM systems.

Figure 4.14 shows the BERs of OFDM and SCM systems against E_b/N_I obtained from the theoretical analysis and computer simulation, and Figure 4.15 shows the BER against the number of subcarriers, $\log_2 N_{SC}$. Here, we assume the GSN with lognormal distribution for $p(\gamma_m)$ and the skewness parameter $B = 2$, which is defined as $B = 20 \log_{10}(E[\gamma_m^2]^{1/2}/E[\gamma_m])$. In addition, we set $E_b/N_0 = 10$ dB and $\lambda_{GS} T = 2.0 \times 10^{-3}$. We calculated $P_b(j)$ by direct integration for $j = 1, 2$, and by Gaussian approximation for $j > 2$. Figure 4.14 also shows the BER lower bound of an OFDM system. We can obtain the lower bound when making the number of subcarriers infinity, which is given by

$$P_b^{N_{SC} \to \infty} = 0.5 \; \text{erfc}\left(\sqrt{E_b/(N_0 + N_I)}\right) \qquad (4.63)$$

In Figures 4.14 and 4.15, the results by the theoretical analysis and by computer simulation are in complete agreement. When the strength of the GSN is comparably high, namely, roughly $E_b/N_I < 10$ dB, the OFDM system cannot outperform the SCM system even if we increase the number of subcarriers. For the moderate strength of the GSN, 10 dB $< E_b/N_I <$ 20 dB, the performance of the OFDM system with a smaller number of subcarriers is worse than that of the SCM system, but the superiority of the OFDM system appears as the number of subcarriers increases. When the strength of the GSN is low, the OFDM system is superior to the SCM system regardless of the number of subcarriers. The main difference in the effect of impulsive noise on the detection process is that impulsive noise interferes with only a few symbols nearby in the SCM system, whereas the energy of

Table 4.2
Transmission Parameters for BER Evaluation in GSN Channels

Modulation/demodulation	BPSK
Guard interval length for OFDM	$\Delta_G/T_S = 0$
Guard interval factor for OFDM	$\alpha_G = 0$
Roll-off factor for SCM	$\alpha_{roll} = 0.25$
Observation window width for SCM	$W_o = 6T$

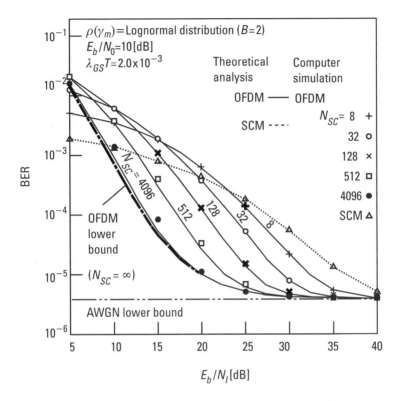

Figure 4.14 BER against E_b/N_I in a lognormal GSN channel.

impulsive noise is spread over all the subcarriers in the OFDM system. As the number of subcarriers increases in the OFDM system, the spread GSN energy at each subcarrier decreases to lead better BER; on the other hand, at the same time, the average number of GSN per OFDM symbol increases to lead a worse BER. When E_b/N_I is high, the DFT can effectively suppress the GSN energy to reduce the number of errors. So, as a result, the BER can be improved as the number of subcarriers increases. When E_b/N_I is low, however, the OFDM system cannot suppress the GSN energy below an appropriately low level, so the DFT operation results in numerous errors over all the subcarriers, while the SCM system can still limit the effect of the GSN within a few symbols.

Figure 4.16 compares the BER lower bound of an OFDM system with the BER of an SCM system. The lower bound decreases slowly in low E_b/N_0 and high N_I/N_0 regions, but it decays very fast when increasing E_b/N_0.

Figure 4.17 shows the BERs of OFDM and SCM systems against the number of subcarriers, which are obtained from the theoretical analysis

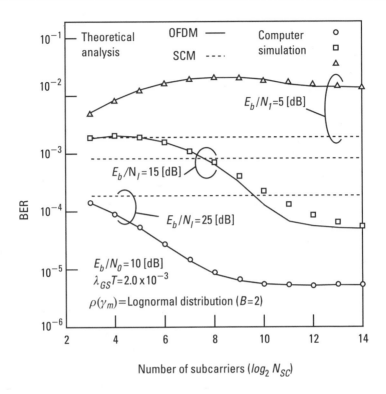

Figure 4.15 BER against a number of subcarriers in a lognormal GSN channel.

assuming the GSN with Laplace distribution for $p(\eta_m)$. The performance in Figure 4.17 is quite similar to one in Figure 4.15, but the slopes of the curves for $E_b/N_I = 15$ dB and 25 dB are a little lower than those in Figure 4.15. This is because the lognormal distribution has a little longer tail than the Laplace distribution.

As shown, an OFDM system is not always more robust to impulsive man-made noises than an SCM system. The BER of an OFDM system highly depends on the number of subcarriers and the strength of the impulsive noises. Our results obtained in this section gives system designers some considerations to choose the appropriate number of subcarriers in an OFDM system so as to obtain the highest robustness against the man-made noise encountered.

4.8 Sensitivity to Frequency Offset

When a frequency offset is introduced in the channel, or when there is a mismatch in the carrier frequency between transmitter and receiver local

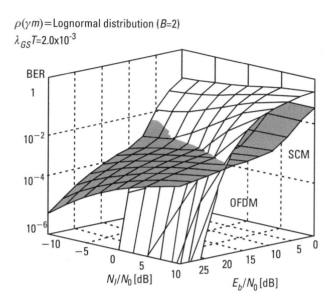

$\rho(\gamma m) =$ Lognormal distribution $(B=2)$

$\lambda_{GS}T = 2.0 \times 10^{-3}$

Figure 4.16 BER lower bound in a lognormal GSN channel.

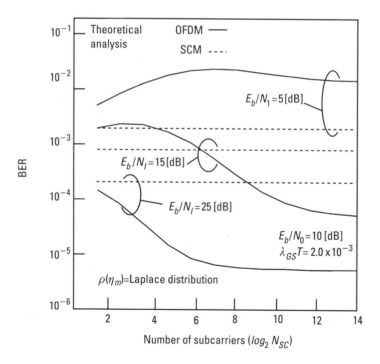

Figure 4.17 BER against a number of subcarriers in a Laplacian GSN channel.

oscillators, the BER degrades drastically, since severe intersubcarrier interference occurs because of the overlapping power spectra between neighboring subcarriers [see Figure 3.8(c)]. This sensitivity to frequency offset is often pointed out as a major OFDM disadvantage.

4.8.1 Bit Error Rate in Frequency Selective and Time Selective Rayleigh Fading Channel with Frequency Offset

Assume that a frequency offset f_{off} and a phase offset ϕ_{off} are introduced in the channel. Then the received signal in equivalent baseband expression is written as

$$r(t) = \int_{-\infty}^{\infty} h(\tau; t)s(t - \tau)e^{-j(2\pi f_{off}t + \phi_{off})}\, d\tau + n(t) \qquad (4.64)$$

Substituting (4.1), (4.30) to (4.32), and (4.64) into (4.49), the nth DFT output is written as

$$r_{ni} =$$

$$\left\{ \sum_{l=1}^{L_1} \frac{c_{ni}}{t_s} e^{-j2\pi f_n \tau_l} \int_{iT_s}^{iT_s+t_s} \alpha_l(t)e^{-j(2\pi f_{off}t + \phi_{off})}\, dt \right.$$

$$+ \sum_{l=L_2+1}^{L_1+L_2} \frac{c_{ni}}{t_s} e^{-j2\pi f_n \tau_l} \int_{iT_s-\Delta_G+\tau_l}^{iT_s+t_s} \alpha_l(t)e^{-j(2\pi f_{off}t + \phi_{off})}\, dt \left. \right\}$$

$$+ \left\{ \sum_{l=1}^{L_1} \sum_{\substack{k=1 \\ k \neq n}}^{N_{SC}} \frac{c_{ki}}{t_s} e^{-j2\pi f_k \tau_l} \int_{iT_s}^{iT_s+t_s} \alpha_l(t)e^{-j2\pi(f_k-f_n)(t-iT_s)}\, dt \right.$$

$$+ \sum_{l=L_1+1}^{L_1+L_2} \sum_{\substack{k=1 \\ k \neq n}}^{N_{SC}} \frac{c_{ki}}{t_s} e^{-jf_k \tau_l} \int_{iT_s-\Delta_G+\tau_l}^{iT_s+t_s} \alpha_l(t)e^{-j2\pi(f_k-f_n)(t-iT_s)}\, dt$$

$$+ \sum_{\substack{l=L_1+1}}^{L_1+L_2} \sum_{\substack{k=1 \\ k \neq n}}^{N_{SC}} \frac{c_{k(i-1)}}{t_s} e^{-j2\pi f_k(\tau_l - T_s)} \int_{iT_s}^{iT_s - \Delta_G + \tau_l} \alpha_l(t) e^{-j2\pi(f_k - f_n)(t - iT_s)} \, dt \Bigg\}$$

$$+ \sum_{l=L_1+1}^{L_1+L_2} \frac{c_{n(i-1)}}{t_s} e^{-j2\pi f_n(\tau_l - T_s)} \int_{iT_s}^{iT_s - \Delta_G + \tau_l} \alpha_l(t) \, dt + n_{ni}$$

$$(4.65)$$

Similarly, as in Section 4.5.2, the components of ρ are written as [28]:

$$\sigma_{S1}^2 = \sum_{l=1}^{L_1} \sigma_l^2 \Bigg[\cos\left(2\pi f_{off} T_s\right) \Bigg\{ \left(\frac{1}{2\pi^2 f_{off}^2 t_s^2} + \frac{3 f_D^2}{4\pi^2 f_{off}^4 t_s^2} - \frac{f_D^2 T_s^2}{2 f_{off}^2 t_s^2} \right) \right.$$

$$\times (1 - \cos\left(2\pi f_{off} t_s\right)) + \frac{f_D^2}{2 f_{off}^2} \cos\left(2\pi f_{off} t_s\right) - \frac{f_D^2}{\pi f_{off}^3 t_s} \sin\left(2\pi f_{off} t_s\right) \Bigg\}$$

$$+ \sin\left(2\pi f_{off} t_s\right)$$

$$\times \left\{ \frac{f_D^2 T_s}{f_{off}^2} \sin\left(2\pi f_{off} t_s\right) - \frac{f_D^2 T_s}{\pi f_{off}^3 t_s^2} (1 - \cos\left(2\pi f_{off} t_s\right)) \right\} \Bigg]$$

$$+ \sum_{l=L_1+1}^{L_1+L_2} \sigma_l^2 \Bigg[\cos\left(2\pi f_{off} T_s\right) \Bigg\{ \left(\frac{1}{2\pi^2 f_{off}^2 t_s^2} + \frac{3 f_D^2}{4\pi^2 f_{off}^4 t_s^2} - \frac{f_D^2 T_s^2}{2 f_{off}^2 t_s^2} \right)$$

$$\times (1 - \cos\left(2\pi f_{off}(t_s + \Delta_G - \tau_l)\right))$$

$$+ \frac{f_D^2 (t_s + \Delta_G - \tau_l)^2}{2 f_{off}^2 t_s^2} \cos\left(2\pi f_{off}(t_s + \Delta_G - \tau_l)\right)$$

$$- \frac{f_D^2 (t_s + \Delta_G - \tau_l)}{\pi f_{off}^3 t_s^2} \sin\left(2\pi f_{off}(t_s + \Delta_G - \tau_l)\right) \Bigg\}$$

$$+ \sin\left(2\pi f_{off} t_s\right) \left\{ \frac{f_D^2 T_s (t_s + \Delta_G - \tau_l)}{f_{off}^2 t_s^2} \times \sin\left(2\pi f_{off}(t_s + \Delta_G - \tau_l)\right) \right.$$

$$- \frac{f_D^2 T_s}{\pi f_{off}^3 t_s} (1 - \cos\left(2\pi f_{off}(t_s + \Delta_G - \tau_l)\right)) \Bigg\} \Bigg]$$

$$(4.66)$$

$$\sigma_{S2}^2 = \sum_{l=1}^{L_1} \sigma_l^2 \left\{ \left(\frac{1}{2\pi^2 f_{off}^2 t_s^2} + \frac{3 f_D^2}{4\pi^2 f_{off}^4 t_s^2} \right) \times (1 - \cos(2\pi f_{off} t_s)) \right.$$

$$\left. + \frac{f_D^2}{2 f_{off}^2} \cos(2\pi f_{off} t_s) - \frac{f_D^2}{\pi f_{off}^3 t_s} \sin(2\pi f_{off} t_s) \right\}$$

$$+ \sum_{l=L_1+1}^{L_1+L_2} \sigma_l^2 \left\{ \left(\frac{1}{2\pi^2 f_{off}^2 t_s^2} + \frac{3 f_D^2}{4\pi^2 f_{off}^4 t_s^2} \right) \right.$$

$$\times (1 - \cos(2\pi f_{off}(t_s + \Delta_G - \tau_l)))$$

$$+ \frac{f_D^2 (t_s + \Delta_G - \tau_l)^2}{2 f_{off}^2 t_s^2} \cos(2\pi f_{off}(t_s + \Delta_G - \tau_l))$$

$$\left. - \frac{f_D^2 (t_s + \Delta_G - \tau_l)}{\pi f_{off}^3 t_s^2} \sin(2\pi f_{off}(t_s + \Delta_G - \tau_l)) \right\}$$

$$+ \sum_{l=L_1+1}^{L_1+L_2} \sigma_l^2 \left\{ \left(\frac{1}{2\pi^2 f_{off}^2 t_s^2} + \frac{3 f_D^2}{4\pi^2 f_{off}^4 t_s^2} \right) \right.$$

$$\times (1 - \cos(2\pi f_{off}(-\Delta_G + \tau_l)))$$

$$+ \frac{f_D^2 (-\Delta_G + \tau_l)^2}{2 f_{off}^2 t_s^2} \cos(2\pi f_{off}(-\Delta_G - \tau_l))$$

$$\left. - \frac{f_D^2 (-\Delta_G + \tau_l)}{\pi f_{off}^3 t_s^2} \sin(2\pi f_{off}(-\Delta_G + \tau_l)) \right\}$$

$$\text{(4.67)}$$

$$\sigma_1^2 = \sum_{l=1}^{L_1} \sum_{\substack{k=1 \\ k \neq n}}^{N_{SC}} \sigma_l^2 \left[\left(\frac{1}{2\pi^2 (f_k - f_n + f_{off})^2 t_s^2} + \frac{3 f_D^2}{4\pi^2 (f_k - f_n + f_{off})^4 t_s^2} \right) \right.$$

$$\times \{1 - \cos(2\pi(f_k - f_n + f_{off}) t_s\}$$

$$+ \frac{f_D^2}{2(f_k - f_n + f_{off})^2} \cos(2\pi(f_k - f_n + f_{off}) t_s)$$

$$- \frac{f_D^2}{\pi(f_k - f_n + f_{off})^3 t_s} \sin(2\pi(f_k - f_n + f_{off})t_s) \Big]$$

$$+ \sum_{l=L_1+1}^{L_1+L_2} \sum_{\substack{k=1 \\ k \neq n}}^{N_{SC}} \sigma_l^2 \left[\left(\frac{1}{2\pi^2(f_k - f_n + f_{off})^2 t_s^2} + \frac{3f_D^2}{4\pi^2(f_k - f_n + f_{off})^4 t_s^2} \right) \right.$$

$$\times \{1 - \cos(2\pi(f_k - f_n + f_{off})(t_s + \Delta_G - \tau_l))\} + \frac{f_D^2(t_s + \Delta_G - \tau_l)^2}{2(f_k - f_n + f_{off})^2 t_s^2}$$

$$\times \cos(2\pi(f_k - f_n + f_{off})(t_s + \Delta_G - \tau_l)) - \frac{f_D^2(t_s + \Delta_G - \tau_l)}{\pi(f_k - f_n + f_{off})^3 t_s^2}$$

$$\times \sin(2\pi(f_k - f_n + f_{off})(t_s + \Delta_G - \tau_l)) \Big]$$

$$+ \sum_{l=L_1+1}^{L_1+L_2} \sigma_l^2 \left[\left(\frac{1}{2\pi^2(f_k - f_n + f_{off})^2 t_s^2} + \frac{3f_D^2}{4\pi^2(f_k - f_n + f_{off})^4 t_s^2} \right) \right.$$

$$\times (1 - \cos(2\pi(f_k - f_n + f_{off})(-\Delta_G + \tau_l))) + \frac{f_D^2(-\Delta_G + \tau_l)^2}{2(f_k - f_n + f_{off})^2 t_s^2}$$

$$\times \cos(2\pi(f_k - f_n + f_{off})(-\Delta_G + \tau_l)) - \frac{f_D^2(-\Delta_G + \tau_l)}{\pi(f_k - f_n + f_{off})^3 t_s^2}$$

$$\times \sin(2\pi(f_k - f_n + f_{off})(-\Delta_G + \tau_l)) \Big]$$

$$(4.68)$$

Substituting (4.35) and (4.66) to (4.68) into (4.28), we can theoretically evaluate the BER of a DPSK-based OFDM system in Rayleigh fading channels with frequency offset.

4.8.2 Numerical Results and Discussions

Figure 4.18 shows the BER of a QDPSK-based OFDM system against the frequency offset normalized by subcarrier separation. Table 4.3 shows the transmission parameters to demonstrate the performance, where the number of subcarriers and the length of guard interval in the OFDM system are

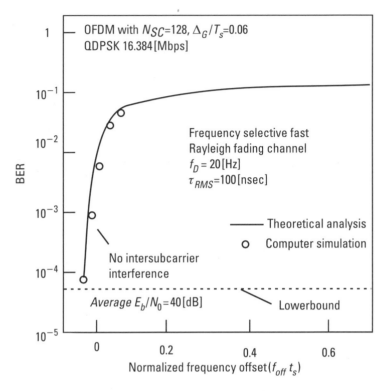

Figure 4.18 BER of QDPSK-based OFDM system against frequency offset.

Table 4.3
Transmission Parameters for BER Evaluation in Fading Channels
with Frequency Offsets

Total symbol transmission rate (R)	8.192 [Msymbols/sec]
Multipath delay profile $(\phi_H(\tau))$	Exponentially decaying 20 paths
RMS delay spread (τ_{RMS})	100 [nsec]
Maximum Doppler frequency (f_D)	20 [Hz]
Modulation/demodulation	QDPSK
Number of subcarriers	$N_{SC} = 128$
Guard interval length	$\Delta_G/T_s = 0.06$

optimized for the transmission and channel parameters in Table 4.1. The theoretical analysis agrees well with the computer simulation result. As compared with the case when ignoring intersubcarrier interference, the OFDM system is very sensitive to the frequency offset; namely, the BER becomes poor even for the small frequency offset. Therefore, we need frequency offset compensation methods to keep good BERs even when a large frequency offset is introduced in the channel. We will discuss this topic in Chapters 5 and 6.

4.9 Sensitivity to Nonlinear Amplification

In general, nonlinear amplifier clips the amplitude of input signal. The sudden changes of the input amplitude generate higher spectral components in the power spectrum of the signal, so it results in a spectrum spreading. The spectrum spreading by nonlinear amplification is a cause of adjacent channel interference (ACI).

Let us consider a case where an OFDM signal is amplified by a nonlinear device. The OFDM signal has nonconstant envelope; in other words, the waveform is like that of a narrowband Gaussian noise because of the central limiting theorem. Therefore, even if we try to reduce the spectrum spreading with a large input back-off, we cannot perfectly eliminate sudden inputs of the larger amplitudes. Furthermore, the outband radiation generated from a subcarrier also becomes "intersubcarrier interference" for the neighboring subcarriers. Therefore, the severe intersubcarrier interference drastically deteriorates the BER performance. This sensitivity to nonlinear amplification is also pointed out as a major OFDM disadvantage, as well as the sensitivity to frequency offset discussed in Section 4.8.

When we can deal with an input signal to a nonlinear amplifier as a narrowband Gaussian noise, we can apply Simbo's method for the theoretical analysis of intermodulation [29].

4.9.1 Simbo's Method

Define the input OFDM signal to a nonlinear amplifier as

$$s_i(t) = N_c(t) + jN_s(t) \tag{4.69}$$

where $N_c(t)$ and $N_s(t)$ are the real and imaginary components of $s_i(t)$, respectively. Furthermore, define the power spectrum of autocorrelation of the input OFDM signal as $W_{OFDM}(f)$ and $R(\tau)$:

$$R(\tau) = \int_{-\infty}^{\infty} W_{OFDM}(f) e^{-j2\pi f \tau} df \tag{4.70}$$

The output of the nonlinear amplifier is given by

$$s_0(t) = g\left(\sqrt{N_c + jN_s}\right) e^{jf\left(\sqrt{N_c + jN_s}\right)} e^{j \tan^{-1}\left(\frac{N_s}{N_c}\right)} \tag{4.71}$$

$$= \int_0^\infty \int_0^\infty \gamma J_1\left(\sqrt{N_c + jN_s}\right) e^{j \tan^{-1}\left(\frac{N_s}{N_c}\right)} J_1(\rho\gamma)\, \rho g(\rho)\, e^{jf(\rho)}\, d\gamma\, d\rho$$

where $g(A)$ and $f(A)$ are the amplitude modulation to amplitude modulation (AM/AM) conversion and amplitude modulation to phase modulation (AM/PM) conversion of the nonlinear amplifier for the input envelope A.

The autocorrelation function of the output signals is written as

$$R_{s_0}(\tau) = \frac{1}{2} E[s_0(t) s_0(t + \tau)] \tag{4.72}$$

Substituting (4.70) and (4.71) into (4.72) leads to:

$$R_{s_0}(\tau) = \frac{1}{4} R(\tau) \sum_{m=0}^\infty \frac{1}{m!(m+1)!} [R(\tau)]^{2m} \tag{4.73}$$

$$\times \left| (-1)^m \int_0^\infty \rho^2 g(\rho)\, e^{jf(\rho)} \left[\frac{\partial^m}{\partial R(0)^m} \left\{ \frac{1}{R^2(0)} e^{-\frac{\rho^2}{2R(0)}} \right\} \right] d\rho \right|^2$$

In (4.73), the term of $m = 0$ corresponds to the (desired) signal component:

$$R_{s_0}^{(s)}(\tau) = R(\tau) \left| \frac{1}{2} \int_0^\infty \rho^2 g(\rho)\, e^{jf(\rho)} \frac{1}{\sigma_s^4} e^{-\frac{\rho^2}{\sigma_s^2}} d\rho \right|^2 \tag{4.74}$$

$$= R(\tau) \left| \frac{1}{2\sigma_s} \int_0^\infty \rho^2 \left\{ g(\sigma_s \rho)\, e^{jf(\sigma,\rho)} \right\} e^{-\frac{\rho^2}{2}} d\rho \right|^2$$

In (4.74), $R(0)$ was replaced by σ_s^2:

$$R(0) = \int_{-\infty}^{\infty} W_{OFDM}(f) \, df = \sigma_s^2 \qquad (4.75)$$

The Fourier transform of (4.74) corresponds to the power spectrum for the signal component:

$$W_o^{(s)}(f) = W_{OFDM}(f) \left| \frac{1}{2\sigma_s} \int_0^{\infty} \rho^2 e^{-\frac{\rho^2}{2}} \left\{ g(\sigma_s \rho) e^{jf(\sigma_s \rho)} \right\} d\rho \right|^2$$

$$(4.76)$$

The term of $m = 1$ in (4.73) corresponds to the third-order intermodulation component:

$$R_{s_o}^{IM3}(\tau) = \frac{1}{8} R(\tau) [R(\tau)]^2 \qquad (4.77)$$

$$\times \left| \frac{1}{\sigma_s^3} \int_0^{\infty} \rho^2 \left(\frac{\rho^2}{2} - 2 \right) e^{-\frac{\rho^2}{2}} \left\{ g(\sigma_s \rho) e^{jf(\sigma_s \rho)} \right\} d\rho \right|^2$$

The power spectrum of the third-order intermodulation is written as

$$W_o^{IM3}(f) = W_{OFDM}(f) \otimes W_{OFDM}(f) \otimes W_{OFDM}(f) \qquad (4.78)$$

$$\times \left| \frac{1}{8\sigma_s^3} \int_0^{\infty} \rho^2 \left(\frac{\rho^2}{2} - 2 \right) e^{-\frac{\rho^2}{2}} \left\{ g(\sigma_s \rho) e^{jf(\sigma_s \rho)} \right\} d\rho \right|^2$$

where $\otimes$ denotes the convolution.

Similarly, as in the above derivations, the autocorrelation function of the fifth-order intermodulation and the power spectrum are written as

$$R_{s_o}^{IM5}(\tau) = \frac{1}{48} R(\tau) \left[R(\tau)\right]^4 \tag{4.79}$$

$$\times \left| \frac{1}{\sigma_s^5} \int_0^\infty \rho^2 \left(\frac{\rho^4}{4} - 3\rho^2 + 6 \right) e^{-\frac{\rho^2}{2}} \left\{ g(\sigma_s \rho) e^{jf(\sigma_s \rho)} \right\} d\rho \right|^2$$

$$W_o^{IM5}(f) = W_{OFDM}(f) \otimes W_{OFDM}(f) \otimes W_{OFDM}(f)$$
$$\otimes W_{OFDM}(f) \otimes W_{OFDM}(f) \tag{4.80}$$

$$\times \left| \frac{1}{\sigma_s^5} \int_0^\infty \rho^2 \left(\frac{\rho^4}{4} - 3\rho^2 + 6 \right) e^{-\frac{\rho^2}{2}} \left\{ g(\sigma_s \rho) e^{jf(\sigma_s \rho)} \right\} d\rho \right|^2$$

4.9.2 Numerical Results and Discussions

To demonstrate some numerical results, we assume a solid state high power amplifier (SSPA) as a nonlinear device. SSPA has a negligibly small AM/PM conversion characteristic, and the AM/AM conversion characteristic is well approximated as [30]

$$g(A) = \frac{A}{(1 + A^{2p})^{(1/2p)}} \tag{4.81}$$

$$f(A) = 0 \tag{4.82}$$

Figure 4.19 shows the AM/AM conversion characteristic in terms of input/output power. As the parameter p increases, the nonlinearity becomes higher. Here we define the input back-off = 0 dB level as the input power with $A = 1$.

For an input OFDM signal to the SSPA, we assume a rectangular power spectrum:

$$W_{OFDM}(f) = \begin{cases} 1, & \left(|f| \leq \frac{1}{2} \right) \\ 0, & \left(|f| > \frac{1}{2} \right) \end{cases} \tag{4.83}$$

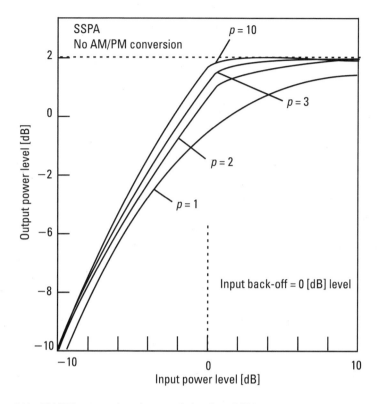

Figure 4.19 AM/AM conversion characteristic of an SSPA.

The three-time and five-time convolutions of $W_{OFDM}(f)$ are written as

$$W_{OFDM}(f) \otimes W_{OFDM}(f) \otimes W_{OFDM}(f) = \qquad (4.84)$$

$$
\begin{cases}
\dfrac{3}{4} - f^2, & \left(|f| \le \dfrac{1}{2}\right) \\[3mm]
\dfrac{1}{2}\left(|f| + \dfrac{3}{2}\right)^2, & \left(\dfrac{1}{2} < |f| \le \dfrac{3}{2}\right) \\[3mm]
0, & \left(|f| > \dfrac{3}{2}\right)
\end{cases}
$$

$$W_{OFDM}(f) \otimes W_{OFDM}(f) \otimes W_{OFDM}(f) \otimes W_{OFDM}(f)$$
$$\otimes\ W_{OFDM}(f) =$$

$$
\begin{cases}
\dfrac{1}{8}\left(f-\dfrac{1}{2}\right)^{4} + \dfrac{1}{3}\left(f-\dfrac{1}{2}\right)^{3} - \dfrac{2}{3}\left(f-\dfrac{1}{2}\right) \\[2mm]
\quad + \dfrac{1}{8}\left(f+\dfrac{1}{2}\right)^{4} - \dfrac{1}{3}\left(f+\dfrac{1}{2}\right)^{3} + \dfrac{2}{3}\left(f-\dfrac{1}{2}\right) & \left(|f| \le \dfrac{1}{2}\right) \\[4mm]
\dfrac{1}{2} - \dfrac{1}{24}\left(f+\dfrac{3}{2}\right)^{4} - \dfrac{1}{8}\left(f+\dfrac{1}{2}\right)^{4} - \dfrac{1}{3}\left(f+\dfrac{1}{2}\right)^{3} \\[2mm]
\quad + \dfrac{2}{3}\left(f+\dfrac{1}{2}\right), & \left(\dfrac{1}{2} < |f| \le \dfrac{3}{2}\right) \\[4mm]
\dfrac{1}{24}\left(f+\dfrac{1}{2}\right)^{4}, & \left(\dfrac{2}{2} < |f| \le \dfrac{5}{2}\right) \\[4mm]
0, & \left(|f| > \dfrac{5}{2}\right)
\end{cases}
$$

$$(4.85)$$

In computer simulation, we assume a BPSK or QDPSK-based OFDM signal with 128 subcarriers. Table 4.4 summarizes the transmission parameters for the nonlinear analysis.

Figures 4.20–4.22 show the power spectra of the OFDM signals amplified by the SSPA for the input back-off = −5, 0, and 5 dB (for the definition of input back-off, see Figure 4.19) [31, 32]. Here, we assume $p = 3$ and take into consideration the intermodulation up to the fifth order. The theoretical results give some underestimation for the power of intermodulation, but they agree with the computer simulation results. As the input back-off increases, the spectrum spreading caused by intermodulation increases.

Table 4.4
Transmission Parameters for Power Spectrum Evaluation by Nonlinear Amplification

Nonlinear amplifier	SSPA ($p = 1,2,3$ and 10)
Modulation/demodulation	QDPSK or BDPSK
Number of subcarriers	$N_{SC} = 128$
Guard interval length	$\Delta_G/T_S = 0$

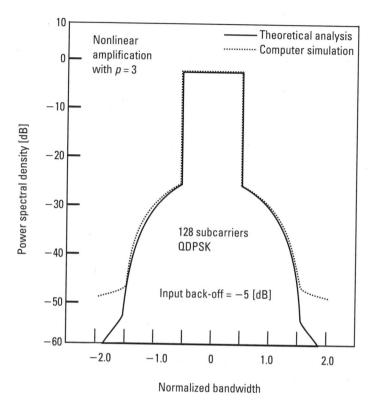

Figure 4.20 Power spectrum of OFDM signal for input back-off −5 dB.

Figure 4.23 shows the ratio of signal power-per-bit to intermodulation power E_b / N_{IM}. Here, we assume no noise and we calculate N_{IM} by integrating (4.77) and (4.79) over the signal bandwidth, namely, $|f| \leq 1/2$. Figure 4.24 shows the BER against the input back-off for the same channel with noise free as in Figure 4.23. As the input back-off increases, the BER drastically degrades because of severe intermodulation. The theoretical result agrees well with the computer simulation result for QDPSK, but there is a large difference between the two results for BDPSK. This may be because we cannot deal with the BDPSK-based OFDM signal as a narrowband Gaussian noise.

We add a Gaussian noise in the channel so as to make the BER be 10^{-4} if a linear amplifier gives the same maximum power as the SSPA (see Figure 4.25). Figure 4.26 shows the BER. As the input back-off increases, the signal power also increases. In the low input back-off range, it makes no intermodulation, so the BER improves. On the other hand, in the high

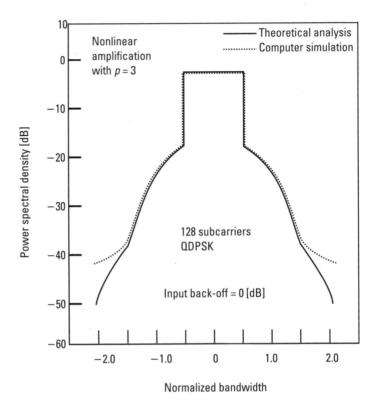

Figure 4.21 Power spectrum of OFDM signal for input back-off 0 dB.

input back-off range, it makes a severe intermodulation, so the BER degrades. Therefore, for a given power of the channel noise, there is an optimum input back-off level to minimize the BER.

Figure 4.27 compares the BERs for different values of p. Here, we assume QDPSK. As the value of p increases, the amplification characteristic improves (see Figure 4.19), so in the low back-off range, the SSPA with a larger p improves the BER. On the other hand, once the input back-off level reaches the optimum value, the SSPA with a larger p deteriorates the BER in the high input back-off range. This is because the larger value of p means higher nonlinearity around the saturation level (see Figure 4.19).

4.10 Sensitivity to A/D and D/A Resolutions

The nonconstant envelope characteristic of an OFDM signal introduces another sensitivity to nonlinear devices such as A/D and D/A converters.

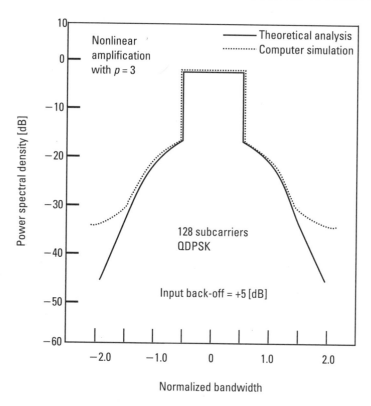

Figure 4.22 Power spectrum of OFDM signal for input back-off +5 dB.

Assume that digital signal processors at the transmitter and receiver have the same q bit-resolution; namely, the transmitter uses a q bit-D/A converter whereas the receiver a q bit-D/A converter. Figure 4.28 shows the system model.

If we can deal with the OFDM signal as a narrowband Gaussian noise with an average of zero and power of σ_n^2, then 99.7% of amplitude values ranges in $[-3\sigma_s, 3\sigma_s]$ and 99.994% of amplitude values ranges in $[-4\sigma_s, 4\sigma_s]$. Here, we call σ_s "effective amplitude." At the transmitter and receiver, nonlinear distortions resulting from quantization and clipping are added to the OFDM signal.

Table 4.5 summarizes the transmission parameters for nonlinear analysis.

Figures 4.29 and 4.30 show the BERs when setting the peak-to-peak quantization range to $6\sigma_s$ and $8\sigma_s$, respectively [32]. For the same value of the A/D and D/A resolution, different quantization ranges give different

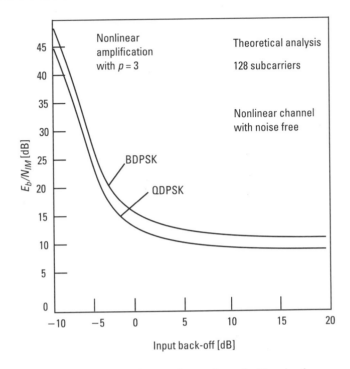

Figure 4.23 Intermodulation power in a nonlinear channel with noise free.

BERs. Figure 4.31 shows the BER against the peak-to-peak quantization range. Note that the BER of 10^{-4} is attainable if using A/D and D/A converters with an infinite bit-resolution. Different resolutions have different optimum values in the quantization ranges to minimize the BER. The 7-bit resolution, with a quantization range of $6\sigma_s$ or $8\sigma_s$, is required to achieve the BER lower bound, namely, the BER of 10^{-4}.

4.11 Conclusions

Every system has both advantages and disadvantages. An OFDM system is robust to frequency selective fading, narrowband interference, and impulsive man-made noises; on the other hand, it is also very sensitive to frequency offset and nonlinear amplification. This chapter discussed the pros and cons of an OFDM system in great detail.

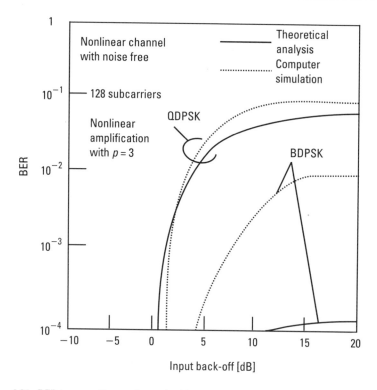

Figure 4.24 BER in a nonlinear channel with noise free.

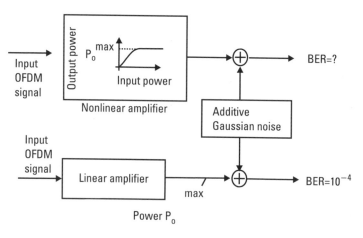

Figure 4.25 Addition of Gaussian noise.

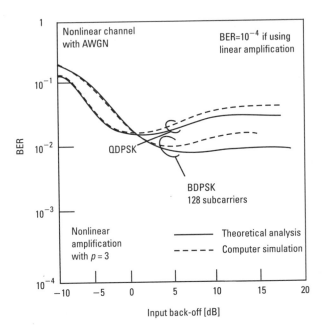

Figure 4.26 BER in a nonlinear channel with AWGN.

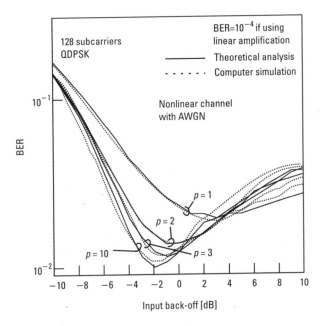

Figure 4.27 BER for different nonlinear characteristics.

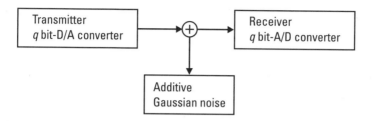

Figure 4.28 Definition of the resolution of A/D and D/A converters.

Table 4.5
Transmission Parameters for BER Evaluation with A/D and D/A Conversions

Resolution of A/D and D/A converters	$q = 3, 4, 5, 6,$ and 7
Peak-to-peak quantization range	$6\sigma_s$ or $8\sigma_s$
Modulation/demodulation	QDPSK or BDPSK
Number of subcarriers	$N_{SC} = 128$
Guard interval length	$\Delta_G/T_S = 0$

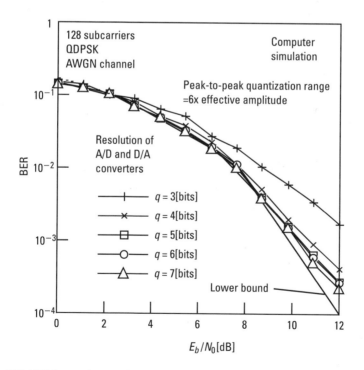

Figure 4.29 BER for peak-to-peak quantization range of $6\sigma_s$.

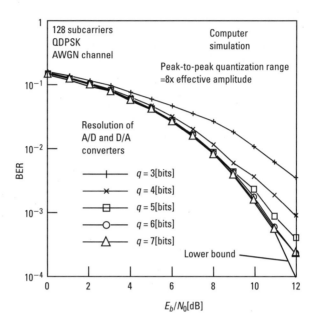

Figure 4.30 BER for peak-to-peak quantization range of $8\sigma_s$.

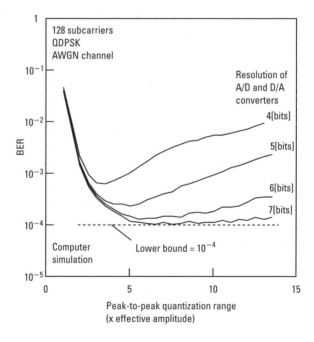

Figure 4.31 BER against peak-to-peak quantization range.

References

[1] van Nee, R., and R. Prasad, *OFDM for Wireless Multimedia Communications,* Norwood, MA: Artech House, 2000.

[2] Proakis, J. G., *Digital Communications, Fourth Edition,* New York: McGraw-Hill, 2001.

[3] Okada, M., S. Hara, and N. Morinaga, "High Speed Indoor Digital Transmission Using Multicarrier Modulation (in Japanese)," *1993 Spring National Conv. Rec. IEICE,* B-394, 1991, pp. 2–394.

[4] Yamane, Y., et al., "Error Performance of High Speed Indoor Transmission Using Multicarrier Modulation Technique (in Japanese)," *IEICE Technical Report,* RCS91-19, 1991, pp. 7–11.

[5] Okada, M., S. Hara, and N. Morinaga, "A Study on Bit Error Rate Performance for Multicarrier Modulation Radio Transmission Systems," *IEICE Technical Report,* RCS91-44, 1991, pp. 19–24.

[6] Okada, M., S. Hara, and N. Morinaga, "Bit Error Rate Performance of Orthogonal Multicarrier Modulation Radio Transmission Systems," *IEICE Trans. on Comm.,* Vol. E76-B, No. 2, February 1993, pp. 113–119.

[7] Hara, S., K. Fukui, and N. Morinaga, "Multicarrier Modulation Technique for Broadband Indoor Wireless Communications," *Proc. IEEEPIMRC'93,* September 1993, pp. 132–136.

[8] Hara, S., M. Okada, and N. Morinaga, "Multicarrier Modulation Technique for Wireless Local Area Networks," *Proc. ECRR'93,* October 1993, pp. 33–38.

[9] Hara, S., and N. Morinaga, "System Performance of Multicarrier Modulation in Fading Channels (in Japanese)," *ITE Technical Report,* ROFT'94-16, Vol. 18, No. 11, February 1994, pp. 39–44.

[10] Hara, S., et al., "Transmission Performance Analysis of Multi-Carrier Modulation in Frequency Selective Fast Rayleigh Fading Channel," *Wireless Personal Communications,* Vol. 2, No. 4, 1995/1996, pp. 335–356.

[11] Okada, M., *High-Speed and High-Quality Transmission Methods for Digital Mobile Communications* (in Japanese), Ph.D. Thesis, Dept. of Comm. Eng., Faculty of Eng., Osaka University, November 1997.

[12] Henkel, W., T. Kessler, and H. Y. Chung, "Coded 64-CAP ADSL in an Impulse-Noise Environment—Modeling of Impulse Noise and First Simulation Results," *IEEE J. Select. Areas Commun.,* Vol. 13, No. 9, December 1995, pp. 1611–1621.

[13] Tsuzuku, A., G. Miyamoto, and Y. Yamanaka, "A Study of the Impulse Noise Characteristics and Its Influence on Digital Broadcasting (in Japanese)," *Proc. IEICE Gen. Conf. '00,* B-5-260, March 2000, p. 645.

[14] Miyamoto, S., and N. Morinaga, "Effect of Microwave Oven Interference on the Performance of Digital Radio Communication Systems," *Proc. IEEE ICC'97,* June 1997, pp. 51–55.

[15] Weinstein, S. B., and P. M. Ebert, "Data Transmission by Frequency-Division Multiplexing Using the Discrete Fourier Transform," *IEEE Trans. Commun. Technol.,* Vol. COM-19, October 1971, pp. 628–634.

[16]   Bingham, J. A. C., "Multicarrier Modulation for Data Transmission: An Idea Whose Time Has Come," *IEEE Commun. Mag.*, Vol. 28, No. 5, May 1990, pp. 5–14.

[17]   Hirosaki, B., "An Orthogonally Multiplexed QAM System Using the Discrete Fourier Transform," *IEEE Trans. Commun.*, Vol. COM-29, No. 7, July 1981, pp. 98–989.

[18]   Chow, J. S., J. C. Tu, and J. M. Cioffi, "A Discrete Multitone Transceiver System for HDSL Applications," *IEEE J. Select. Areas Commun.*, Vol. 9, No. 6, August 1991, pp. 895–908.

[19]   Kosmopoulos, S. A., et al., "Fourier-Bessel Error Performance Analysis and Evaluation of M-ary QAM Schemes in an Impulsive Noise Environment," *IEEE Trans. Commun.*, Vol. COM-39, No. 3, March 1991, pp. 398–404.

[20]   Bello, P. A., and R. Esposito, "A New Method for Calculating Probabilities of Errors Due to Impulsive Noise," *IEEE Trans. Commun. Tech.*, Vol. COM-17, No. 6, June 1969, pp. 368–379.

[21]   Ghosh, M., "Analysis of the Effect of Impulsive Noise on Multicarrier and Singlecarrier QAM Systems," *IEEE Trans. Commun.*, Vol. COM-44, No. 2, February 1996, pp. 145–147.

[22]   Budsabathon, M., and S. Hara, "Robustness of OFDM Signal Against Temporally Localized Impulsive Noise," *Proc. IEEE VTC 2001-Fall*, Atlantic City, NJ, October 7–11, 2001, available in CD-ROM.

[23]   Budsabathon, M., and S. Hara, "Robustness of OFDM System Against Temporally Localized Man-Made-Noises," *IEICE Trans. Fundamentals*, Vol. E85-A, No. 10, October 2002, pp. 2336–2344.

[24]   Mertz, P., "Model of Impulsive Noise for Data Transmission," *IEEE Trans. Commun. Sys.*, Vol. CS-9, No. 6, June 1961, pp. 130–137.

[25]   Lind, L.F., and N. A. Mufti, "Efficient Method for Modeling Impulse Noise in a Communication System," *IEE Electron. Lett.*, Vol. 32, August 1996, pp. 1440–1441.

[26]   Papoulis, A., *Probability, Random Variables and Stochastic Processes*, New York: McGraw-Hill, 1991.

[27]   Helstrom, C. W., "Calculating Error Probabilities for Intersymbol and Cochannel Interference," *IEEE Trans. Commun.*, Vol. COM-34, No. 5, May 1986, pp. 430–435.

[28]   Fukui, K., *Frequency Tracking Method for Multicarrier Modulation Systems* (in Japanese), Master Thesis, Dept. of Comm. Eng., Faculty of Eng., Osaka University, February 1994.

[29]   Simbo, O., *Transmission Analysis in Communication Systems Volume 2*, New York: Computer Science Press, 1988.

[30]   Rapp, C., "Effects of HPA-Nonlinearity on a 4-DPSK/OFDM Signal for a Digital Sound Broadcasting System," *Proc. 2nd European Conference on Satellite Communications*, October 1991, pp. 179–184.

[31]   Mouri, M., S. Hara, and N. Morinaga, "Effect of Nonlinear Amplification on Orthogonal Multicarrier Modulation (in Japanese)," *1994 Spring National Conv. Rec.*, B-445, 1994, pp. 2–445.

[32]   Mouri, M., *Effect of Nonlinear Amplification on Orthogonal Multicarrier Modulation* (in Japanese), Bachelor Thesis, Dept. of Comm. Eng., Faculty of Eng., Osaka University, February 1994.

Appendix 4A

Define a (2×1) column vector as

$$\mathbf{r} = [r_{ni}, r_{n(i-1)}]^T \tag{4A.1}$$

where r_{ni} and $r_{n(i-1)}$ are the complex-valued Gaussian random variables with zero average and normalized power, and T denotes the transpose. The vector $\mathbf{r}$ has the following (2×2) covariance matrix:

$$\mathbf{R} = \frac{1}{2} E[\mathbf{r}^* \mathbf{r}^T] = \begin{bmatrix} 1 & \rho \\ \rho^* & 1 \end{bmatrix} \tag{4A.2}$$

On the other hand, the decision variable $f = 2D_{ni}$ is written as

$$f = 2D_{ni} = 2 \operatorname{Re}\left[r_{ni} r^*_{n(i-1)} e^{j\left(\frac{\pi}{2} - \frac{\pi}{M}\right)} \right] = \mathbf{r}^H \mathbf{F} \mathbf{r} \tag{4A.3}$$

where $\mathbf{F}$ is given by the following (2×2) matrix:

$$\mathbf{F} = \begin{bmatrix} 0 & e^{j\left(\frac{\pi}{2} - \frac{\pi}{M}\right)} \\ e^{-j\left(\frac{\pi}{2} - \frac{\pi}{M}\right)} & 0 \end{bmatrix} \tag{4A.4}$$

and H denotes the Hermitian transpose. The characteristic function, which is defined as the Fourier transform of the p.d.f. of d, is given by [1]

$$C_d(\eta) = \int_{-\infty}^{\infty} e^{j\eta f} p(f) \, df \tag{4A.5}$$

$$= \frac{1}{\det(\mathbf{I} - 2j\eta \mathbf{R}^* \mathbf{F})}$$

where $\det(*)$ denotes the determinant of matrix $*$ and $\mathbf{I}$ is the (2×2) identity matrix. Therefore, $p(f)$ can be obtained by the inverse Fourier transform of $C_d(\eta)$ as

$$p(f) = \frac{1}{2\pi} \int_{-\infty}^{\infty} e^{-jnf} C_d(\eta) \, d\eta \qquad (4A.6)$$

$$= \begin{cases} \dfrac{B}{\sqrt{A^2 + 4B}} e^{\frac{1}{2}\left(A + \sqrt{A^2 + 4B}\right)f}, & (f \le 0) \\[3mm] \dfrac{B}{\sqrt{A^2 + 4B}} e^{\frac{1}{2}\left(A - \sqrt{A^2 + 4B}\right)f}, & (f > 0) \end{cases}$$

where A and B are given by

$$A = \frac{\text{Re}\left[\rho e^{-j\left(\frac{\pi}{2} - \frac{\pi}{M}\right)}\right]}{1 - |\rho|^2} \qquad (4A.7)$$

$$B = \frac{1}{1 - |\rho|^2} \qquad (4A.8)$$

Finally, we can obtain (4.28) by integrating $p(f)$ over the error region:

$$P_{b,\,fading}^{B\,and\,Q,\,differential} = \Pr\{f = 2D_{ni} \le 0\} = \int_{-\infty}^{0} p(f) \, df \qquad (4A.9)$$

Reference

[1]   Schwartz, M., W. R. Bennett, and S. Stein, *Communication Systems and Techniques*, New York: IEEE Press, 1996.

Appendix 4B

The magnitude of the normalized correlation is given by (4.29). Using (4.33), $E[r_{ni} r_{n(i-1)}^*]$ and $E[r_{ni} r_{ni}^*]$ in (4.29) are written as

$$E[r_{ni} r^*_{n(i-1)}] = \sum_{l=1}^{L_1} \frac{1}{t_s^2} \int_{0}^{t_s} \int_{-T_s}^{-iT_s+t_s} \phi_{H,l}(x-y)\, dx\, dy \tag{4B.1}$$

$$+ \sum_{l=L_1+1}^{L_1+L_2} \int_{-\Delta_G+\tau_l}^{t_s} \int_{-T_s-\Delta_G+\tau_l}^{-T_s+t_s} \phi_{H,l}(x-y)\, dx\, dy$$

$$E[r_{ni} r^*_{ni}] = \sum_{l=1}^{L_1} \frac{1}{t_s^2} \int_{0}^{t_s} \int_{0}^{t_s} \phi_{H,l}(x-y)\, dx\, dy$$

$$+ \sum_{l=L_1+1}^{L_1+L_2} \int_{-\Delta_G+\tau_{l_s}}^{t_s} \int_{-\Delta_G+\tau_l}^{t_s} \phi_{H,l}(x-y)\, dx\, dy$$

$$+ \sum_{l=1}^{L_1} \sum_{\substack{k=1 \\ k \neq n}}^{N_{SC}} \frac{1}{t_s^2} \int_{0}^{t_s} \int_{0}^{t_s} \phi_{H,l}(x-y)\, e^{j2\pi(f_k-f_n)(x-y)}\, dx\, dy \tag{4B.2}$$

$$+ \sum_{l=L_1+1}^{L_1+L_2} \sum_{\substack{k=1 \\ k \neq n}}^{N_{SC}} \frac{1}{t_s^2} \int_{-\Delta_G+\tau_l}^{t_s} \int_{-\Delta_G+\tau}^{t_s} \phi_{H,l}(x-y)\, e^{j2\pi(f_k-f_n)(x-y)}\, dx\, dy$$

$$+ \sum_{l=L_1+1}^{L_1+L_2} \sum_{k=1}^{N_{SC}} \frac{1}{t_s^2} \int_{0}^{-\Delta_G+\tau_l} \int_{0}^{-\Delta_G+\tau_l} \phi_{H,l}(x-y)\, e^{j2\pi(f_k-f_n)(x-y)}\, dx\, dy$$

$$+ \sigma_n^2$$

where $\phi_{H,l}(\Delta t)$ is the autocorrelation function of $\alpha_l(t)$, which is defined as [see (2.22)]

$$\phi_{H,l}(\Delta t) = \frac{1}{2} E[\alpha_l(t + \Delta t)\, \alpha_l^*(t)] \tag{4B.3}$$

For the Jakes' model with an omnidirectional antenna, $\phi_{H,l}(\Delta t)$ is given by (2.33). Furthermore, when $f_D \Delta t \ll 1$, it can be approximated as

$$\phi_{H,l}(\Delta t) = \sigma_l^2 J_0(2\pi f_D \Delta t) = \sigma_l^2 \{1 - (\pi f_D \Delta t)^2\} \qquad (4B.4)$$

Substituting (4B.4) into (4B.1) and (4B.2), ρ, which is defined by (4.29), becomes (4.35).

5

Pilot-Assisted DFT Window Timing/ Frequency Offset Synchronization and Subcarrier Recovery

5.1 Introduction

Synchronization, which is composed of estimation and control, is one of the most important functionalities of the receiver. It must be first performed by the receiver when receiving information data.

In an OFDM system, synchronization can be divided into three different parts—carrier frequency offset synchronization, DFT window timing synchronization, and subcarrier recovery. As shown in Section 4.8, an OFDM system is very sensitive to frequency offset, which may be introduced in the radio channel, so accurate frequency offset synchronization is essential. Especially for burst mode data transmission in wireless LAN applications, we must keep the overhead, namely, the number of pilot symbols required for the synchronization, as low as possible. DFT window timing synchronization corresponds to symbol timing synchronization in single carrier transmission. However, it is much more difficult, because there is always an "eye opening" in each single carrier modulated symbol, whereas there are many "zero crossings" in each OFDM symbol. Therefore, normal synchronization algorithms such as zero-forcing cannot be adopted. Furthermore, subcarrier recovery means simultaneous regeneration of reference signals at all subcarriers used.

This chapter discusses pilot-assisted DFT window timing/frequency offset synchronization and subcarrier recovery methods suited for burst mode OFDM data transmission. Accurate synchronization often requires accurate estimation, so some estimation methods are also investigated in this chapter. Section 5.2 introduces Schmidl's DFT window timing/frequency offset estimation method [1] and presents the theoretical analysis and computer simulation results. Section 5.3 discusses two DFT window timing synchronization/subcarrier recovery methods, which periodically insert time domain pilot (TDP) symbols or frequency domain pilot (FDP) symbols into a train of OFDM data symbols [2–6]. Section 5.4 presents a pilot symbol generation method suited for an OFDM signal. Section 5.5 concludes the topic.

Figure 5.1 shows a whole system model that is composed of a chain of an OFDM transmitter, a radio channel, and an OFDM receiver. This model is used to discuss the performance of three methods for DFT window timing estimation/synchronization, frequency offset estimation or subcarrier recovery. For the radio channel, we assume an AWGN channel, a static 20-path channel, a static 30-path channel, or a frequency selective Rayleigh fading channel. Through the radio channel, an unknown frequency offset f_{off} and an unknown time delay δ_d are introduced. Frequency offset may result from a mismatch of local oscillator frequency between transmitter and receiver, but we can include its contribution into the frequency offset introduced in the radio channel. On the other hand, for the time delay introduced in the radio channel, we set $\delta_d = 0$ in this chapter. Therefore, a task of DFT window timing synchronizer is to find $\delta_d = 0$.

Figure 5.2 shows the structure of an OFDM transmitter. A pilot signal composed of a known sequence is inserted into a train of OFDM signals in a time division manner. On the other hand, Figure 5.3 shows the structure of an OFDM receiver. With assistance from the transmitted pilot signal, DFT window timing synchronization, frequency offset compensation, and subcarrier recovery are performed before data demodulation. We will discuss the details in Section 5.2 and 5.3. Note that, in Figure 5.3, f_c denotes a

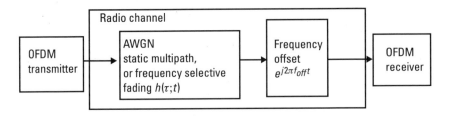

Figure 5.1 Pilot-assisted system model.

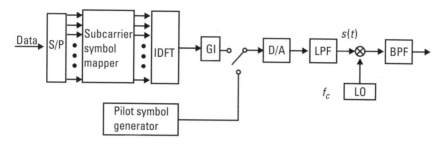

Figure 5.2 OFDM transmitter model (TDP type).

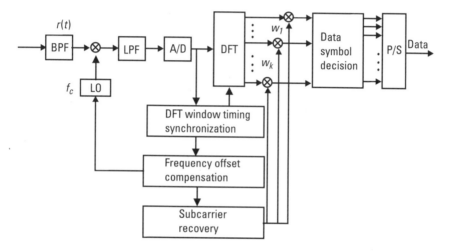

Figure 5.3 OFDM receiver model.

carrier frequency. The reason to show the radio frequency is that we just need to adjust the frequency of local oscillator (LO) to compensate for the frequency offset. Therefore, it has no special reason, and we will carry out analysis in the equivalent baseband form, as we have done in previous chapters.

5.2 Pilot-Assisted DFT Window Timing/Frequency Offset Estimation Method

5.2.1 Principle of DFT Window Timing Estimation

Figure 5.4 shows a signal burst format for Schmidl's method, where the preamble is 1 OFDM symbol long and the data are composed of several

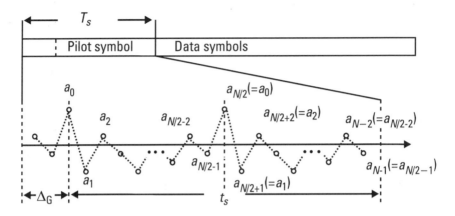

Figure 5.4 Schmidl's signal burst format.

OFDM symbols. Now we will call the preamble excluding the guard interval "the pilot symbol." The pilot symbol is composed of N known samples and has two identical halves in the time domain. The pilot symbol is written as

$$a_m = \begin{cases} a(mt_s/N), & (m = 0, 1, \ldots, N/2 - 1) \\ a_{m-N/2}, & (m = N/2, N/2 + 1, \ldots, N - 1) \end{cases} \quad (5.1)$$

Assuming an AWGN channel (see Figure 5.1), the sampled received signal is written as

$$r_m = r(mt_s/N) = s_m + n_m = a_m e^{j2\pi \frac{F_{off}}{N} m} + n_m \quad (5.2)$$

where $F_{off} (= f_{off} t_s)$ is the frequency offset normalized by the subcarrier separation, and s_m and n_m are the signal and noise samples, respectively.

Define the correlation between the received signal and its $t_s/2$-delayed version as follows:

$$R(d) = \sum_{m=0}^{N/2-1} r_{d+m}^* r_{d+m+N/2} \quad (5.3)$$

where d is a time index corresponding to the first sample in a DFT window. As previously mentioned, in the pilot symbol, the first half is identical to the second half, therefore, when $d = 0$, the magnitude of the correlation will be a large value. This implies that (5.3) can be used as a measure to find a DFT window timing:

find $\hat{d}$ which maximizes $M(d)$

$$M(d) = m(d)^2 = \frac{|R(d)|^2}{S(d)^2} \tag{5.4}$$

where $S(d)$ is the received energy for the second half symbol given by

$$S(d) = \sum_{m=0}^{N/2-1} |r_{d+m+N/2}|^2 \tag{5.5}$$

Here, we call $M(d)$ "the timing metric," and the fact that $M(d)$ takes the maximal value at $d = 0$ results in correct estimation of $\delta_d = 0$.

Figure 5.5 shows the timing metric of an OFDM signal in an AWGN channel. Here, Table 5.1 summarizes the transmission parameters. In this case, the timing metric reaches a plateau that has a length equal to the length of the guard interval (it is 51 samples wide in the figure), and the start of

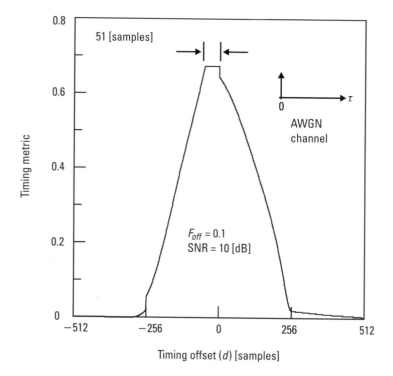

Figure 5.5 Timing metric for an AWGN channel.

Table 5.1
Transmission Parameters for Evaluation of Timing Metric

Number of subcarriers	512
Guard interval length	$\Delta_G/t_s = 0.1$ (51 [samples])
Channel model	SNR = 10 dB
	AWGN,
	Static equal gain 20-path
	(No path beyond guard interval),
	Static equal gain 30-path
	(10 paths beyond guard interval)

a DFT window can be taken to be anywhere within this plateau. On the other hand, Figure 5.6 shows the timing metric in the static 20-path channel, where all the paths have the same gain and there is no path beyond the guard interval (see the impulse response in the same figure). In this case, the length of plateau is equal to that of the guard interval minus that of the channel impulse response interval (it is 32 samples wide in the figure).

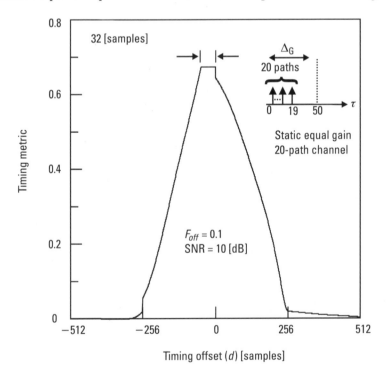

Figure 5.6 Timing metric for a static 20-path (no path beyond guard interval).

Therefore, the plateau is shorter than for the AWGN channel. Furthermore, Figure 5.7 shows the timing metric in the static 30-path channel, where all the paths have the same gain and there are 10 paths beyond the guard interval (see the impulse response in the same figure). In this case, there is no distinct plateau any more because of severe ISI.

5.2.2 Principle of Frequency Offset Estimation

Furthermore, in the pilot symbol, the same sample is transmitted after time interval $t_s/2$, so the frequency offset can be estimated with the angle of (5.3) as (see Figure 5.8)

$$\hat{F}_{off} = \frac{1}{\pi} \angle R(\hat{d}) = \frac{1}{\pi} \tan^{-1} \left[\frac{\text{Im}\{R(\hat{d})\}}{\text{Re}\{R(\hat{d})\}} \right] \tag{5.6}$$

When no ISI occurs in calculation of (5.6), this method can give a good estimate for the frequency offset. Therefore, as long as the frequency

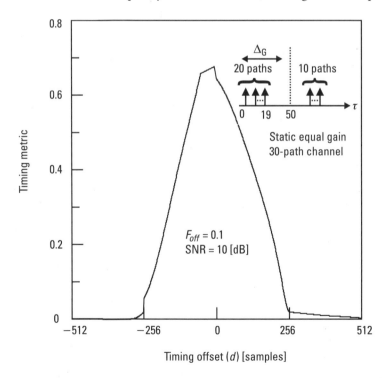

Timing offset (d) [samples]

Figure 5.7 Timing metric for a static 30-path (10-path beyond guard interval).

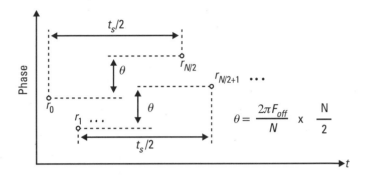

Figure 5.8 Phase shift caused by frequency offset.

offset is calculated near the best timing point, this method is valid even for frequency selective fading channels. Figure 5.9 shows the block diagram of the DFT window timing/frequency offset estimation method.

5.2.3   Spectral Property of Pilot Symbol

In general, DFT deals with an observed windowed waveform as a period of a periodic waveform. Therefore, N-point DFT of the pilot symbol, with observation window width of t_s, could give N spectral components with frequency resolution of $1/t_s$. However, the pilot symbol has two identical halves with period of $t_s/2$, so it can have spectral components at integer

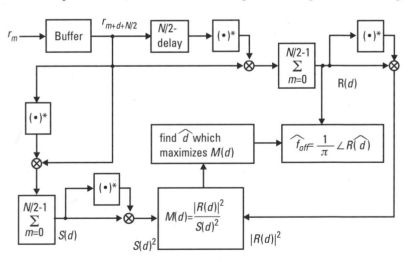

Figure 5.9 Block diagram of Schmidl's DFT window timing/frequency offset estimation method.

multiples of $2/t_s$. This means that, in the frequency spectrum of the pilot symbol obtained through N-point DFT, there are nonzero components at even frequency indexes (total $N/2$ spectral components) and zeros at odd frequency indexes. Figure 5.10 shows the spectral property of the pilot symbol. This spectral property introduces a wider frequency estimation range up to $1/t_s$, which is equal to the subcarrier separation and is twice as wide as Moose's method [7].

5.2.4 Performance of DFT Window Timing Estimator

As defined by (4.1), the OFDM signal is a sum of many signals with different subcarrier frequencies, so we can assume that the inphase and quadrature components are Gaussian by the central limiting theorem. Now, we define the powers of the signal and noise in (5.2) as follows:

$$E[\text{Re}\{s_m\}^2] = E[\text{Im}\{s_m\}^2] = \sigma_s^2 \tag{5.7}$$

$$E[\text{Re}\{n_m\}^2] = E[\text{Im}\{n_m\}^2] = \sigma_n^2 \tag{5.8}$$

therefore, the SNR is given by σ_s^2/σ_n^2.

Assume an optimal DFT timing, namely, $\hat{d} = 0$. In this case, $R(0)$ is written as

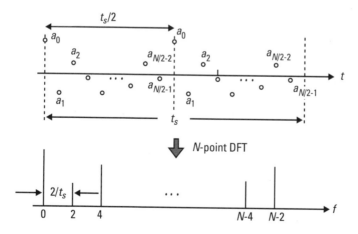

Figure 5.10 Spectral property of Schmidl's pilot symbol.

$$R(0) = \sum_{m=0}^{N/2-1} \left\{ |a_m|^2 \, e^{j2\pi \frac{F_{off}}{N} \frac{N}{2}} + a_m^* \, e^{j2\pi \frac{F_{off}}{N} m} \, n_{m+N/2} \right.$$

$$\left. + a_m^* \, e^{j2\pi \frac{F_{off}}{N} \left(m + \frac{N}{2}\right)} \, n_{m+N/2} + n_m^* \, n_{m+N/2} \right\} \tag{5.9}$$

In (5.9), the first term is dominant, so (5.9) has the angle of πF_{off} from the inphase axis [this is the reason why we can estimate the frequency offset from (5.3)]. For the sake of calculation, it is convenient to multiply $R(0)$ by $e^{-j\pi F_{off}}$ to make the angle zero:

$$R(0) e^{-j\pi F_{off}} = \sum_{m=0}^{N/2-1} \left\{ |a_m|^2 + a_m^* \, e^{j2\pi \frac{F_{off}}{N} \left(m - \frac{N}{2}\right)} \, n_{m+N/2} \right.$$

$$\left. + a_m^* \, e^{j2\pi \frac{F_{off}}{N} m} \, n_{m+N/2} + n_m^* \, n_{n+N/2} \, e^{-j\pi F_{off}} \right\} \tag{5.10}$$

When the SNR is high, the fourth term in (5.10) is negligibly small, so it means that we can deal with $R(0) e^{-j\pi F_{off}}$ as a complex-valued Gaussian random variable [this multiplication does not change the magnitude of $R(0)$]:

$$E[\operatorname{Re}\{R(0) e^{-j\pi F_{off}}\}] = N\sigma_s^2 \tag{5.11}$$

$$E[\operatorname{Im}\{R(0) e^{-j\pi F_{off}}\}] = 0 \tag{5.12}$$

$$E[\operatorname{Re}\{R(0) e^{-j\pi F_{off}}\}^2 - (N\sigma_s^2)^2] = 2N\sigma_s^2 \sigma_n^2 \tag{5.13}$$

$$E[\operatorname{Im}\{R(0) e^{-j\pi F_{off}}\}^2] = 2N\sigma_s^2 \sigma_n^2 \tag{5.14}$$

Furthermore, as compared with the inphase component, the quadrature component is small and can be neglected, so the magnitude of $R(0)$ is given by

$$|R(0)| = n(N\sigma_s^2, \, 2N\sigma_s^2 \sigma_n^2) \tag{5.15}$$

where $n(\mu, \sigma^2)$ denotes a Gaussian random variable with average of μ and variance of σ^2.

On the other hand, $S(0)$ is written as

$$S(0) = \sum_{m=0}^{N/2-1} \left\{ |a_m|^2 + |n_{m+N/2}|^2 \right\} \tag{5.16}$$

$$+ \sum_{m=0}^{N/2-1} \left\{ a_m^* e^{-j2\pi \frac{F_{off}}{N} m} n_{m+N/2} + a_m e^{j2\pi \frac{F_{off}}{N} m} n_{m+N/2}^* \right\}$$

$S(0)$ is a real-valued Gaussian random variable, but we can approximate it as a constant, because the standard deviation is much smaller than the average:

$$E[S(0)] = N(\sigma_s^2 + \sigma_n^2) \tag{5.17}$$

Therefore, $m(0)$ defined by (5.4) is written as

$$m(0) = \frac{|R(0)|}{S(0)} = \frac{n(N\sigma_s^2, 2N\sigma_s^2 \sigma_n^2)}{N(\sigma_s^2 + \sigma_n^2)}$$

$$= n\left(\frac{\sigma_s^2}{\sigma_s^2 + \sigma_n^2}, \frac{2\sigma_s^2 \sigma_n^2}{N(\sigma_s^2 + \sigma_n^2)^2} \right)$$

$$= n\left(\frac{\text{SNR}}{\text{SNR} + 1}, \frac{2\text{SNR}}{N(\text{SNR} + 1)^2} \right) \tag{5.18}$$

$$\cong n\left(1, \frac{2}{N \cdot \text{SNR}} \right)$$

$$= 1 + n\left(0, \frac{2}{N \cdot \text{SNR}} \right)$$

namely, $m(0)$ is a Gaussian random variable with an average of 1 and variance of $2/(N \cdot \text{SNR})$. Consequently, $M(0)$ is written as

$$M(0) = m(0)^2 = \left\{ 1 + n\left(0, \frac{2}{N \cdot \text{SNR}} \right) \right\}^2$$

$$= 1 + 2n\left(0, \frac{2}{N \cdot \text{SNR}} \right) + n\left(0, \frac{2}{N \cdot \text{SNR}} \right)^2 \tag{5.19}$$

$$\cong 1 + n\left(0, \frac{8}{N \cdot \text{SNR}} \right)$$

$$= n\left(1, \frac{8}{N \cdot \text{SNR}} \right)$$

Equation (5.19) shows that $M(0)$ is a Gaussian random variable with an average of 1 and variance of $8/(N \cdot SNR)$.

5.2.5 Performance of Frequency Offset Estimator

With (5.10), where $R(0)$ is rotated by πF_{off}, (5.6) is rewritten as (also with $\hat{d} = 0$)

$$\hat{F}_{off} = F_{off} + \frac{1}{\pi} \tan^{-1} \left[\frac{\text{Im}\{R(0)\,e^{-j2\pi F_{off}}\}}{\text{Re}\{R(0)\,e^{-j2\pi F_{off}}\}} \right] \qquad (5.20)$$

For small argument $x \ll 1$, we can approximate $\tan^{-1}(x) \cong x$, so (5.20) is simplified to

$$\hat{F}_{off} \cong F_{off} + \frac{1}{\pi} \frac{\text{Im}\{R(0)\,e^{-j2\pi F_{off}}\}}{\text{Re}\{R(0)\,e^{-j2\pi F_{off}}\}} \qquad (5.21)$$

The real part of $R(0)\,e^{-j\pi F_{off}}$ is a real-valued Gaussian random variable, but we can approximate it as a constant, because the standard deviation is much smaller than the average. Therefore, (5.21) leads to:

$$\hat{F}_{off} = F_{off} + \frac{1}{\pi} \left\{ 0 + \frac{n(0,\, 2N\sigma_s^2\,\sigma_n^2)}{N\sigma_s^2} \right\}$$

$$= n\left(F_{off},\, \frac{2N\sigma_s^2\,\sigma_n^2}{\pi^2 N^2 \sigma_s^4} \right) \qquad (5.22)$$

$$= n\left(F_{off},\, \frac{2}{\pi^2 N \cdot SNR} \right)$$

namely, the estimate of frequency offset is a Gaussian random variable with an average of F_{off} and variance of $2/(\pi^2 N \cdot SNR)$.

Figure 5.11 shows the frequency estimation performance in the three different channels introduced in Section 5.2.1. Here, the best DFT window timing is assumed ($\delta_d = 0$). The performance in the AWGN and static 20-path channels is better than that in the static 30-path channel. Even if each path has a time variation, this method can give a good estimate for frequency offset as long as the length of the channel impulse response is less than that of the guard interval and the best DWT window timing is found.

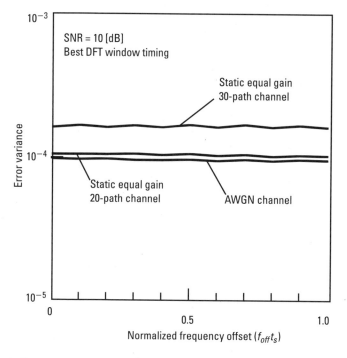

Figure 5.11 Frequency offset estimation performance.

5.3 Pilot-Assisted DFT Window Timing Synchronization and Subcarrier Recovery Method

5.3.1 Time Domain Pilot-Assisted DFT Window Timing Synchronization and Subcarrier Recovery Method

The previous method correlates the received signal with its delayed version to perform DFT window timing/frequency offset estimation. The method works well, although it requires a shorter pilot length, but it also requires an additional pilot symbol to estimate instantaneous impulse response or instantaneous frequency response of channel essential for subcarrier recovery.

This section introduces a method that also inserts a pilot symbol into a train of OFDM symbols in a time division manner as shown in Figure 5.2. The method can perform DFT window timing synchronization and subcarrier recovery, but it cannot perform frequency offset estimation.

Figure 5.12 shows the structure of the data frame and the pilot waveform for the TDP method, where a pilot symbol is inserted in every N_t OFDM symbol. The pilot symbol is composed of a baseband pulse-shaped pseudo

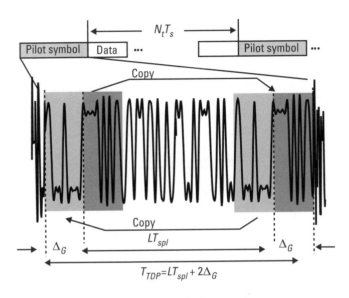

Figure 5.12 Structure of a TDP data frame and pilot waveform.

noise (PN) sequence based on a maximum length shift register code, and it is L symbol long with length of LT_{spl}. If a Nyquist filter with roll-off factor of α_{roll} is used for the pulse shaping, to meet the requirement of the same bandwidth, the roll-off factor must satisfy the following condition [see (3.20)]:

$$B_{Nyquist} = (1 + \alpha_{roll})\,R = \frac{R}{1 - \alpha_G} \qquad (5.23)$$

therefore, we have

$$\alpha_{roll} = \frac{\alpha_G}{1 - \alpha_G} \qquad (5.24)$$

In Figure 5.12, to eliminate ISI from neighboring signals, the PN sequence is extended in its head and tail parts, so the pilot symbol length is given by

$$T_{TDP} = LT_{spl} + 2\Delta_G \qquad (5.25)$$

Note that we will discuss how to generate a pilot symbol when there is a restriction on its length and bandwidth in Section 5.4.

The OFDM signal is transmitted through the frequency selective fading channel with impulse response of $h(\tau; t)$ (see Figure 5.1). Figure 5.13 shows the block diagram of the DFT window timing synchronization and subcarrier recovery for the TDP method. The pilot symbol part is first fed into the matched filter to estimate the impulse response $\tilde{h}(\tau; t)$, and then the best DFT window timing is examined through finding the maximum in the estimated impulse response.

Assume that we have just estimated a channel impulse response at receiver clock $t = 0$ and the path at $\tau = \tau_m$ has the largest gain among all the paths within the guard interval Δ_G. Figure 5.14 shows the estimated impulse response. In this case, the DFT window timing is set to $t_w = \tau_m + \Delta_G$. Here, we assume that the guard interval is composed of N_{spl} samples

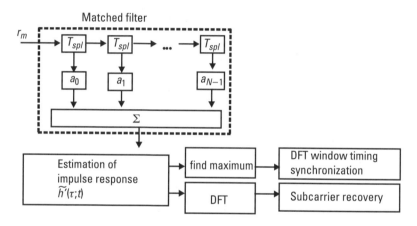

Figure 5.13 Block diagram of TDP method.

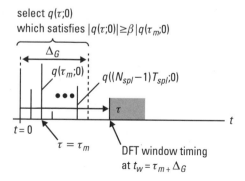

Figure 5.14 Estimation criterion on channel impulse response.

and we can have exactly N_{spl} estimated path gains. The impulse response is given by

$$\bar{h}(\tau; 0) = \sum_{l=0}^{N_{spl}-1} q(lT_{spl}; 0)\,\delta(\tau - lT_{spl}) \qquad (5.26)$$

where $q(lT_{spl}; 0)$ is the estimated gain for the lth path. Among the estimated path gains, especially the weaker gains, there may be some wrong ones caused by noise. To eliminate such wrong gains, we set a threshold as

$$\bar{h}'(\tau; 0) = \sum_{l=0}^{N_{spl}-1} q'(lT_{spl}; t_m)\,\delta(\tau - lT_{spl}) \qquad (5.27)$$

$$q'(lT_{spl}; 0) = \begin{cases} q(lT_{spl}; 0), & \left(|q(lT_{spl}; 0)| \geq \beta\,|q(\tau_m; 0)|\right) \\ 0, & \text{(otherwise)} \end{cases}$$
$$(5.28)$$

where β is a path selection threshold.

Now we have an estimated impulse response, so we can obtain a frequency response essential for subcarrier recovery through its DFT. The complex-valued envelope for the kth subcarrier is given by

$$\bar{H}(f_k; t_w) = \sum_{n=0}^{N_{SC}-1} \bar{h}'(\tau = nT_{spl}; 0)\,e^{-j2\pi n f_k} \qquad (5.29)$$

where

$$h'(\tau = nT_{spl}; 0) = 0, \qquad (N_{spl} \leq n \leq N_{SC} - 1) \qquad (5.30)$$

Finally, the weight for the kth subcarrier recovery (coherent demodulation) is given by

$$w_k = \frac{\bar{H}^*(f_k, t_w)}{|\bar{H}(f_k, t_w)|^2} \qquad (5.31)$$

Note that the weights for subcarrier recovery can be obtained only once in the data interval composed of N_t OFDM symbols. Therefore, to track the time variation of the channel and to produce reference signals over the OFDM data interval, we use a linear interpolation of the obtained weights in the time domain.

5.3.2 Frequency Domain Pilot-Assisted Subcarrier Recovery Method

Figure 5.15 shows the transmitter block diagram for the FDP method, where known pilot symbols are inserted in frequency/time division manner [8, 9]. Figure 5.16 shows the frequency/time signal format. A pilot symbol is inserted in every N_f subcarrier in the frequency domain and in every N_t OFDM symbol in the time domain. Figure 5.17 shows an interpolation method using pilot symbols to estimate frequency responses at data subcarriers. In the time domain, linear interpolation is also used to cope with time variation of channel.

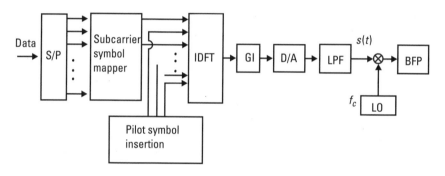

Figure 5.15 OFDM transmitter model (FDP type).

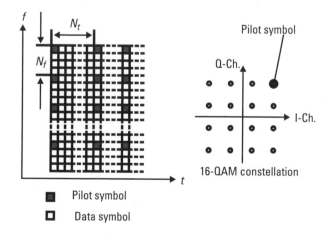

Figure 5.16 Frequency/time signal format.

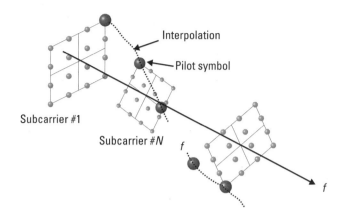

Figure 5.17 Interpolation in frequency domain.

5.3.3 Numerical Results and Discussions

Table 5.2 shows the transmission parameters to demonstrate the BER performance of the TDP and FDP methods.

Figure 5.18 shows the BER of the TDP method in an AWGN channel. For both $L = 63$ and 511, selection of a larger β gives a better BER. This is because, for the AWGN channel, there is only one real path in the estimated impulse response, and a smaller β increases wrong selections of paths caused by noise. Therefore, setting a larger β improves the BER. On the other hand, the BER for $L = 63$ is superior to that for $L = 511$. The PN sequence with $L = 511$ has a better autocorrelation property, but it has a longer length.

Table 5.2
Transmission Parameters for BER Evaluation

Total symbol transmission rate (R)	16.348 [Msymbols/sec]
Number of subcarriers	512
Guard interval length	$\Delta_G/t_s = 0.1$ (51 [samples])
Modulation/Demodulation	CQPSK, 16-QAM
Length of PN sequence	$L = 63, 511$ (BPSK)
	Maximum length shift register sequence
Roll-off factor for pilot symbol	$\alpha_{roll} = 0.11$
OFDM symbol interval for pilot	$N_t = 10$
Subcarrier interval for pilot	$N_f = 2, 4, 8,$ and 16
Time domain interpolation	Linear
Frequency domain interpolation	Cubic spline, polynomial, and linear
Channel model	AWGN, 2-path i.i.d.,
	6-path exponentially decaying

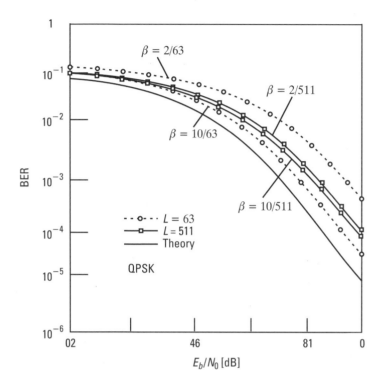

Figure 5.18 BER of TDP method in an AWGN channel.

In the computer simulation, the signal energy is also allocated for the pilot symbol, so the energy loss associated with the longer pilot insertion is dominant, as compared with the improvement in the autocorrelation property.

Figure 5.19 shows the BER versus the path selection threshold in a frequency selective fast Rayleigh fading channel with a 6-path exponentially decaying multipath delay profile. Setting a smaller β increases the probability that paths caused by noise are wrongly selected, whereas setting a larger β increases the probability that real paths are wrongly *not*-selected. Therefore, for a given E_b/N_0, there is an optimum value in the path selection threshold to minimize the BER. From the figure, $\beta = 0.1$ and $\beta = 0.5$ are proper choices for $L = 511$ and $L = 63$, respectively. In the following figures, we set $\beta = 0.1$ for $L = 511$ and $\beta = 0.5$ $L = 63$, respectively.

Figure 5.20 shows the error variance of the recovered reference signal versus the RMS delay spread normalized by the DFT window width for the FDP method, where a frequency selective fast Rayleigh fading channel with a 6-path exponentially decaying multipath delay profile is assumed. In general,

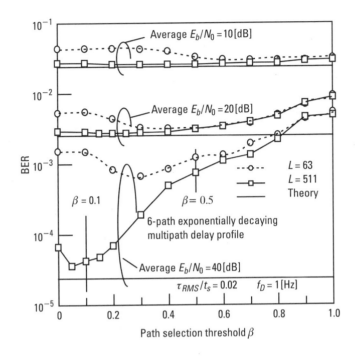

Figure 5.19 BER versus path selection threshold for TDP method.

as the normalized RMS delay spread increases, the error variance increases. The performance of the polynomial interpolation method is worse because of the wild oscillation between the tabulated points (pilot symbols). The cubic spline interpolation method performs best among the three methods, and the performance of $N_f = 8$ is almost the same as that of $N_f = 4$. In the following figures, we use the cubic spline interpolation method.

Figures 5.21 and 5.22 show the BER versus the normalized RMS delay spread for frequency selective fast Rayleigh fading channels with 2-path i.i.d. multipath delay profile and 6-path exponentially decaying multipath delay profile, respectively. In the two figures, we set noise free. The FDP method can perform well in the region of smaller delay spread, but the BER becomes worse as the delay spread increases. This is because the wider coherence bandwidth results in accurate estimation of the frequency response by the cubic spline interpolation method when the delay spread is small, whereas the narrower coherence bandwidth introduces a larger estimation error when the delay spread is large. On the other hand, the performance of the TDP method is relatively flat for variation of the delay spread and it largely depends on the length of the PN sequence selected.

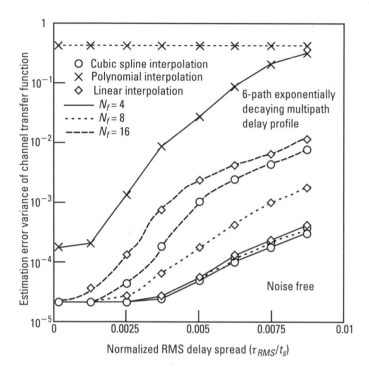

Figure 5.20   Error variance versus RMS delay spread for FDP method.

Figure 5.23 shows the BER versus the maximum Doppler shift normalized by the pilot insertion interval in a frequency selective fast Rayleigh fading channel with 2-path i.i.d. multipath delay profile, where we assume 16 quadrature amplitude modulation (QAM). In addition, here we set noise free. For $f_D T_{plt} < 0.1$, the estimation error in the channel impulse response or channel transfer function is dominant, as compared with the tracking error of the channel time variation, so the performance of the TDP ($L = 511$) and FDP ($N_f = 2$) methods is superior to that of the TDP ($L = 63$) and FDP ($N_f = 4$) methods. On the other hand, for $f_D T_{plt} > 0.1$, where the channel tracking error is dominant, there is no large difference among the four curves and they become worse as the normalized maximum Doppler shift increases.

Figures 5.24 and 5.25 show the BER versus the average E_b/N_0 in frequency selective fast Rayleigh fading channels with 2-path i.i.d. multipath delay profile and 6-path exponentially decaying multipath delay profile, respectively. In the two figures, the theoretical BER of 16 QAM for flat fading is given by [9]

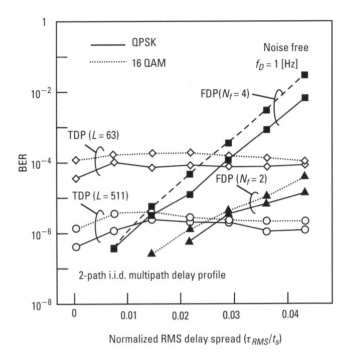

Figure 5.21 BER versus RMS delay spread (2-path i.i.d. multipath delay spread).

$$P_{b, fading}^{16QAM, coherent} = \frac{3}{8}\left(1 - \sqrt{\frac{4(1 - \alpha_G)\overline{\gamma_b}}{10 + 4(1 - \alpha_G)\overline{\gamma_b}}}\right) \qquad (5.32)$$

Note that, in the computer simulation, the normalized delay spread is uniformly distributed in [0.001, 0.04]. Therefore, some events have narrower coherence bandwidths and others have wider coherence bandwidths. For the FDP method with $N_f = 4$, the BER shows a high BER floor. This is because it cannot correctly estimate the channel transfer function regardless of the coherence bandwidth. For the FDP method with $N_f = 2$, the BER shows no BER floor, but there is a penalty in average E_b/N_0 from the theoretical lower bound. This is because it cannot correctly estimate the channel transfer function by way of the cubic spline interpolation when the coherence bandwidth is narrower. On the other hand, the TDP method with $L = 511$ shows no BER floor and the performance is very close to the theoretical lower bound, although the TDP method with $L = 63$ shows a BER floor because of its bad autocorrelation property.

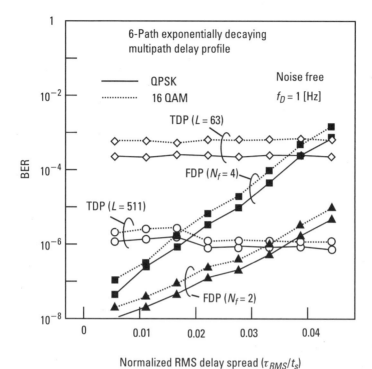

Figure 5.22 BER versus RMS delay spread (6-path exponentially decaying multipath delay spread).

5.4 Chaotic Pilot Symbol Generation Method

In general, subcarrier recovery requires a PN sequence as a known pilot symbol to estimate the channel response. The PN sequence adopted in Section 5.3 was based on a baseband pulse-shaped maximum length shift register code. It has a good autocorrelation property, but it has some restriction, namely, the length should be $2^Q - 1$ samples, where Q is an integer. Therefore, when there is a restriction on its length and also its required bandwidth, we cannot adopt this approach. We need to look for an alternative for PN sequence generation.

Figure 5.26 shows a PN sequence generation method [10], where a PN sequence is first generated in the frequency domain. Here, we use a chaotic method using the following logistic map:

$$x_{n+1} = 4x_n(1 - x_n) \tag{5.33}$$

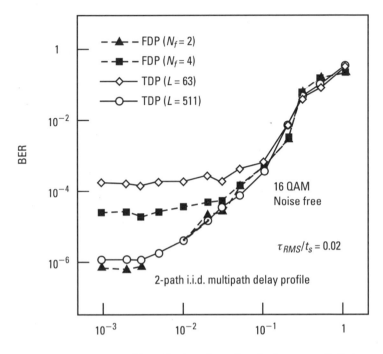

- - ▲ - - FDP (N_f = 2)

- - ■ - - FDP (N_f = 4)

—◇— TDP (L = 63)

—○— TDP (L = 511)

16 QAM
Noise free

τ_{RMS}/t_s = 0.02

2-path i.i.d. multipath delay profile

BER

Normalized maximum Doppler frequency ($f_D T_{plt}$)

Figure 5.23 BER versus maximum Doppler frequency.

Using the logistic map, we can have a random sequence uniformly distributed in [0, 1.0] with infinite length. To map an obtained random variable to one of the QPSK signal constellations a_p, we use the following map:

$$a_p = \begin{cases} 00, & (0 \leq x_n < 0.25) \\ 01, & (0.25 \leq x_n < 0.5) \\ 10, & (0.5 \leq x_n < 0.75) \\ 11, & (0.75 \leq x_n < 1.0) \end{cases} \qquad (5.34)$$

Now, we have a PN sequence in the frequency domain, so then we can have a PN sequence as a pilot symbol in the time domain by way of its DFT and cyclic extension. In Figure 5.26, we first generate a frequency domain-PN sequence spanned over 52 subcarriers [Figure 5.26(a)], and we finally have a pilot symbol composed of 80 samples, through 64-point IFFT and 16 sample-cyclic extension [Figure 5.26(b)]. Figure 5.27 shows the

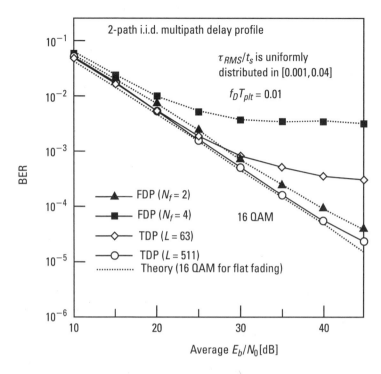

Figure 5.24 BER versus average E_b/N_0 (2-path i.i.d. multipath delay spread).

autocorrelation property of a generated pilot symbol. The pilot symbol has a relatively good autocorrelation property. The merit of this method is that there is no restriction on the length and bandwidth of the obtained PN sequence and that we can generate a lot of PN sequences to check the autocorrelation and peak to average power ratio (PAPR) properties.

5.5 Conclusions

We introduced Schmidl's method for DFT window timing/frequency offset estimation in Section 5.2. It requires just an OFDM symbol-long pilot symbol, but it shows good estimation performance. We confirmed it in the theoretical analysis and computer simulation results. However, to carry out subcarrier recovery essential for coherent demodulation, Schmidl's method requires an additional pilot symbol.

We discussed two methods for DFT window timing synchronization/ subcarrier recovery, the TDP type and the FDP type. Our computer

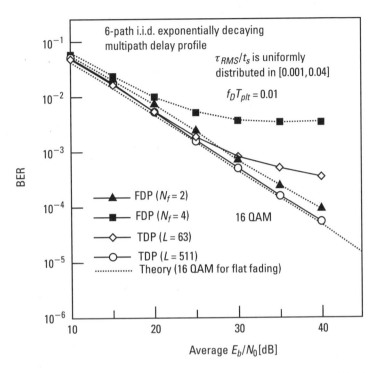

Figure 5.25 BER versus average E_b/N_0 (6-path exponentially decaying multipath delay spread).

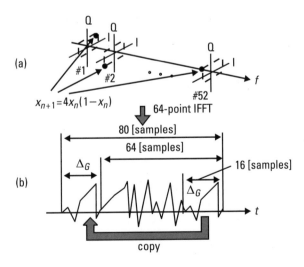

Figure 5.26 Chaotic PN sequence generation: (a) PN sequence generation in frequency domain; and (b) pilot symbol generation in time domain.

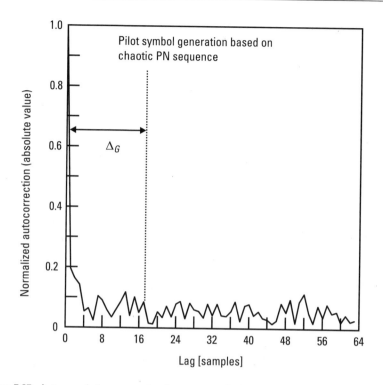

Figure 5.27 Autocorrelation property of a generated pilot symbol.

simulation results show that the TDP method is more robust to the variation of delay spread.

Finally, we introduced a chaotic pilot symbol generation method suited for an OFDM signal. The method uses the logistic map to generate a random sequence with infinite length and can release a restriction on the symbol length obtained.

References

[1] Schmidl, T. M., and D. C. Cox, "Robust Frequency and Timing Synchronization for OFDM," *IEEE Trans. Commun.,* Vol. COM-45, No. 12, December 1997, pp. 1613–1621.

[2] Imamura, D., S. Hara, and N. Morinaga, "A Spread Spectrum-Based Subcarrier Recovery Method for Orthogonal Multi-Carrier Modulated Signal (in Japanese)," *IEICE Technical Report,* RCS97-116, 1997, pp. 21–27.

[3] Imamura, D., S. Hara, and N. Morinaga, "A Study on Pilot Signal Aided Subcarrier Recovery Method for Orthogonal Multi-Carrier Modulated Signal (in Japanese)," *IEICE Technical Report,* RCS97-222, 1997, pp. 69–76.

[4]   Imamura, D., S. Hara, and N. Morinaga, "A Spread Spectrum-Based Subcarrier Recovery Method for Orthogonal Multi-Carrier Modulation System (in Japanese)," *Proc. IEICE Gen. Conf. '98,* B-5-15, 1998, p. 379.

[5]   Imamura, D., *A Study on Pilot Signal Aided Subcarrier Recovery Method for Orthogonal Multi-Carrier Modulated Signal* (in Japanese), Master Thesis, Dept. of Comm. Eng., Faculty of Eng., Osaka University, February 1998.

[6]   Imamura, D., S. Hara, and N. Morinaga, "Pilot-Assisted Subcarrier Recovery Methods for OFDM Systems," *IEICE Trans. on Commun.,* Vol. J82-B, No. 3, March 1999, pp. 292–401.

[7]   Moose, P. H., "A Technique for Orthogonal Frequency Division Multiplexing Frequency Offset Correction," *IEEE Trans. Commun.,* Vol. COM-42, No. 10, October 1994, pp. 2908–2914.

[8]   Yamashita, I., S. Hara, and N. Morinaga, "A Pilot Signal Insertion Technique for Multicarrier Modulation using 16 QAM (in Japanese)," *1994 IEICE Spring National Conv. Rec.,* B-356, 1994, pp. 2–356.

[9]   Hara, S., et al., "Transmission Performance Analysis of Multi-Carrier Modulation in Frequency Selective Fast Rayleigh Fading Channel," *Wireless Personal Communications,* Vol. 2, No. 4, 1995/1996, pp. 335–356.

[10] Hara, S., S. Hane, and Y. Hara, "Does OFDM Really Prefer Frequency Selective Fading Channels?" *Multi-Carrier Spread-Spectrum and Related Topics,* Dordrecht: Kluwer Academic Publishers, 2002, pp. 35–42.

6

Blind Maximum Likelihood-Based Joint DFT Window Timing/ Frequency Offset/DFT Window Width Estimation

6.1 Introduction

In Chapter 5, we discussed some pilot-assisted DFT window timing and frequency offset estimation methods and a pilot-assisted subcarrier recovery method. In this chapter, we discuss a blind or pilotless joint DFT window timing, frequency offset, and DFT window width estimation method based on a maximum likelihood criterion.

Several blind parameter estimation methods have been proposed for OFDM systems [1, 2]. As compared with a pilot-assisted approach, in general, a blind approach requires longer observation symbols to give accurate estimates for parameters we want to estimate. On the other hand, as shown in Chapters 3 and 4, to maintain orthogonality among subcarriers even in multipath fading channels, each OFDM symbol is cyclically extended with a guard interval, whose waveform is exactly the same as the tail of the symbol itself [see Figure 3.8(a)]. In other words, an OFDM transmitter transmits the same waveform twice in each symbol period that is still unknown, but we can deal with it as "an unknown pilot signal," unlike a normal pilot signal whose waveform we know. This waveform structure introduces the cyclostationary property of an OFDM signal [3]. Making effective use of that very property, we can obtain a good estimate within shorter observation symbols [4, 5].

This chapter is organized as follows. A system model is presented in Section 6.2. Section 6.3 analyzes the maximum likelihood parameter estimation for cyclostationary signal. Numerical results and discussions are summarized in Section 6.4. Finally, conclusions are given in Section 6.5.

6.2 System Model

Figure 6.1 shows a system model to discuss the estimation performance. Through the channel, the transmitted signal is perturbed by an unknown time delay δ_d, an unknown impulse response $h(\tau; t)$, and an unknown frequency f_{off}. The frequency offset changes the OFDM symbol width, so the receiver needs to estimate the DFT window timing δ_d, the frequency offset f_{off}, and the DFT window width t_s before data demodulation.

The received signal is written as

$$r(t) = (h \otimes s)(t - \delta_d) e^{j2\pi f_{off} t} + n(t) \tag{6.1}$$

where $s(t)$ is given by (4.1), and $n(t)$ is an AWGN with power spectral density of $N_0/2$ (see (4.7)). $(h \otimes s)(t)$ denotes the convolution of $h(\tau; t)$ and $s(t)$, which is given by

$$(h \otimes s)(t) = \int_{-\infty}^{+\infty} h(\tau; t) s(t - \tau) \, d\tau \tag{6.2}$$

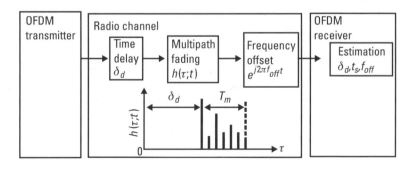

Figure 6.1 Blind system model.

Assuming that the channel is WSSUS, then from (2.17), the autocorrelation function of the channel is given by

$$R_h(u, v) = \phi_h(u, v; 0)$$

$$= \frac{1}{2} E[h^*(u; t) h(v; t)] \qquad (6.3)$$

$$= g(u)\, \delta(u - v)$$

where $g(t)$ is the multipath delay profile $(= \phi_h(t))$ defined by (2.19). Here, we assume that $h(\tau; t)$ or $g(t)$ has a support only over $0 < t \le T_m$ (see Figure 6.1). Therefore, we need to set Δ_G to be more than T_m.

6.3 Maximum Likelihood Parameter Estimation for Cyclostationary Signal

Consider a problem to estimate δ_d, f_{off} and t_s by observing $r(t)$ in the observation period $[0, M_o T_s]$, where M_o denotes the number of observation symbols.

When $s(t)$, $h(\tau; t)$, δ_d, f_{off} and t_s are given, the conditional p.d.f. of $r(t)$ is given by

$$p(r|s, h, \delta_d, f_{off}, t_s) = A e^{-X} \qquad (6.4)$$

where A is a constant and X is the Euclidian distance between a known transmitted signal and the received signal:

$$X = \frac{1}{N_0} \int_{t \in I} |r(t) - (h \otimes s)(t - \delta_d) e^{j(2\pi f_{off} t + \theta)}|^2 \, dt \qquad (6.5)$$

We could obtain the likelihood function for δ_d, f_{off} and t_s by averaging (6.4) in terms of $s(t)$ and $h(\tau; t)$, but it could be difficult. Therefore, we first expand (6.4) in Taylor series, then in the obtained series, we select a few terms for averaging, which contribute to the parameter estimation. We can obtain the likelihood function as follows:

$$\Lambda(\delta_d, f_{off}, t_s) = \int\limits_{u \in I} \int\limits_{v \in I} r(u) R_{hs}(u - \delta_d, v - \delta_d) \qquad (6.6)$$

$$\times r^*(v) e^{j2\pi f_{off}(u-v)} \, du \, dv$$

where $R_{hs}(u, v)$ is the autocorrelation function of $(h \otimes s)(t)$:

$$R_{hs}(u, v) = \frac{1}{2} E[(h \otimes s)^*(u)(h \otimes s)(v)] \qquad (6.7)$$

$$= \int\limits_{-\infty}^{\infty} \int\limits_{-\infty}^{\infty} R_h(\xi, \eta) R_s(u - \xi, v - \eta) \, d\xi \, d\eta$$

In (6.7), $R_s(u, v)$ is the autocorrelation function of the transmitted signal:

$$R_s(u, v) = \frac{1}{2} E[s^*(u) s(v)] \qquad (6.8)$$

Substituting (6.3) into (6.7) leads to:

$$R_{hs}(u, v) = \int\limits_{0}^{T_m} g(\xi) R_s(u - \xi, v - \xi) \, d\xi \qquad (6.9)$$

The transmitted signal $s(t)$ has a cyclostationary property and the autocorrelation function is written as

$$R_s(u, v) = \sum_{m=-\infty}^{\infty} R_s'(u - mT_s, v - mT_s) \qquad (6.10)$$

where

$$R_s'(u, v) = \begin{cases} \sum\limits_{l=1}^{N_{SC}} \sum\limits_{k=1}^{N_{SC}} \frac{1}{2} E[c_{km}^* c_{lm}] e^{j2\pi(ku-lv)/t_s}, & \begin{array}{l} (-\Delta_G < u \le t_s, \\ -\Delta_G < v \le t_s) \end{array} \\ 0, & \text{(otherwise)} \end{cases}$$

$$(6.11)$$

and c_{km} is a complex-valued random variable with average 0 and variance 1:

$$\frac{1}{2}E[c^*_{km}c_{km}] = \begin{cases} 1, & (k = l) \\ 0, & (\text{otherwise}) \end{cases} \tag{6.12}$$

Therefore, $R'_s(u, v)$ can be rewritten as

$$R'_s(u, v) = \begin{cases} \displaystyle\sum_{k=1}^{N_{SC}} e^{j2\pi k(u-v)/t_s}, & (-\Delta_G < u \le t_s, -\Delta_G < v \le t_s) \\ 0, & (\text{otherwise}) \end{cases} \tag{6.13}$$

Furthermore, when the number of subcarriers is large, $N_{SC} \gg 1$, $R'_s(u, v)$ can be approximated as

$$R'_s(u, v) = \begin{cases} \delta(u - v) + \delta(u - v + t_s) \\ + \delta(u - v - t_s), & (-\Delta_G < u \le t_s, -\Delta_G < v \le t_s) \\ 0, & (\text{otherwise}) \end{cases} \tag{6.14}$$

so from (6.9), (6.10), and (6.14), $R_{hs}(u, v)$ can be written as

$$R_{hs}(u, v) = B\delta(u - v) + \sum_{m=-\infty}^{\infty} R'_{hs}(u - mT_s, v - mT_s) \tag{6.15}$$

where B is a constant and

$$R'_{hs}(u, v) = g_r(v)\,\delta(u - v - t_s) + g_r(u)\,\delta(u - v + t_s) \tag{6.16}$$

$$g_r(t) = \begin{cases} \displaystyle\int_0^{\min(\Delta_G+t, T_m)} g(\xi)\,d\xi; & (-\Delta_G < t \le 0) \\[2mm] \displaystyle\int_t^{\min(\Delta_G+t, T_m)} g(\xi)\,d\xi; & (0 < t \le T_m) \\[2mm] 0; & (\text{otherwise}) \end{cases} \tag{6.17}$$

By substituting (6.15) into (6.6), we can calculate the likelihood function as

$$
\Lambda(\delta_d, f_{off}, t_s) = B \int_{t \in I} |r(t)|^2 \, dt
$$

$$
+ 2 \, \mathrm{Re}\left[\sum_{m=1}^{M} \int_{-\Delta_G}^{T_m} g_r(t) r(t + \delta_d - mT_s) \right. \qquad (6.18)
$$

$$
\left. \times \, r^*(t + t_s + \delta_d - mT_s) \, dt e^{j2\pi f_{off} t_s} \right]
$$

However, in (6.18), the first term does not contribute to the parameter estimation, so we can redefine the likelihood function as

$$
\lambda(\delta_d, f_{off}, t_s) = \mathrm{Re}\left[\sum_{m=1}^{M} \int_{-\Delta_G}^{T_m} g_r(t) r(t + \delta_d - mT_s) \right. \qquad (6.19)
$$

$$
\left. \times \, r^*(t + t_s + \delta_d - mT_s) \, dt e^{j2\pi f_{off} t_s} \right]
$$

Equation (6.19) still contains an unknown parameter $g_r(t)$, which we cannot estimate before the parameter estimation. However, if the delay spread is not so large, we can approximate it as

$$
g_r(t) \cong \begin{cases} 1, & (-\Delta_G < t \leq 0) \\ 0, & (\text{otherwise}) \end{cases} \qquad (6.20)
$$

Therefore, we can finally obtain the likelihood function as

$$\lambda(\delta_d, f_{off}, t_s) = \mathrm{Re}\left[e^{j2\pi f_{off} t_s} \sum_{m=1}^{M} \int_{-\Delta_G}^{0} r(t + \delta_d - mT_s) \right. \quad (6.21)$$

$$\left. \times r^*(t + t_s + \delta_d - mT_s)\, dt \right]$$

We can estimate δ_d, f_{off} and t_s by searching for their values that maximize (6.21). Figure 6.2 shows the block diagram of the estimator. We call it "the optimum estimator" in the maximum likelihood sense.

The optimum estimator contains an integrator, so it is somewhat complicated. We can obtain "the suboptimum estimator" when dropping the integration in (6.21):

$$\lambda'(\delta_d, f_{off}, t_s) = \mathrm{Re}\left[e^{j2\pi f_{off} t_s} \sum_{m=1}^{M} r(t + \delta_d - mT_s) \right. \quad (6.22)$$

$$\left. \times r^*(t + t_s + \delta_d - mT_s) \right]$$

Figure 6.3 shows the suboptimum estimator.

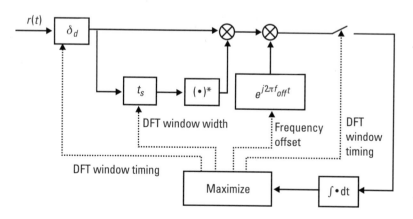

Figure 6.2 Optimum estimator.

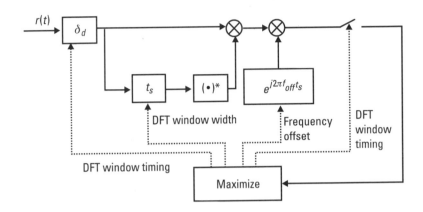

Figure 6.3 Suboptimum estimator.

6.4 Numerical Results and Discussions

We evaluate the RMS DFT window timing error, RMS frequency error, RMS DFT window width error, and BER obtained through 5,000 computer simulation runs. Table 6.1 summarizes the transmission parameters to evaluate the estimation performance. Here, we employ a DPSK signal format to avoid subcarrier recovery.

Figures 6.4, 6.5, 6.6, and 6.7 show the RMS DFT window timing error, RMS frequency error, RMS DFT window width error, and BER in an AWGN channel. Here, the frequency error and DFT window width are normalized by T_s, whereas the frequency error is by t_s. The optimum estimator can give accurate estimates for δ_d, f_{off}, and t_s and can achieve a good BER within 10 to 20 observation symbols. To obtain the BER close

Table 6.1
Transmission Parameters for Evaluation of Estimation Performance

Number of subcarriers	128
Modulation/Demodulation	QDPSK
Total symbol transmission rate (R)	8.192 [Msymbols/sec]
Symbol duration (T_s)	12.2 [μsec]
Guard interval (Δ_G)	350 [nsec]
Multipath delay profile ($\phi_H(\tau)$)	Exponentially decaying 10 paths
Envelope distribution	Rayleigh
RMS delay spread (τ_{RMS})	100 [nsec]
Time selectivity	Slow

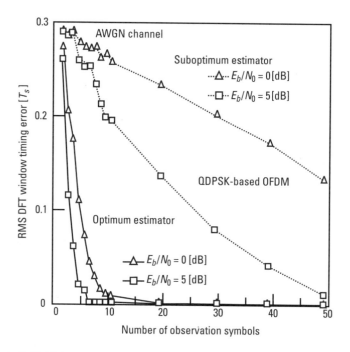

Figure 6.4 RMS DFT window timing error in an AWGN channel.

to the lower bound, the optimum estimator requires 10 observation symbols, whereas the suboptimum estimator requires 50 observation symbols.

Figures 6.8, 6.9, 6.10, and 6.11 show the RMS DFT window timing error, RMS frequency error, RMS DFT window width error, and BER in a frequency selective Rayleigh fading channel. The optimum estimator can also give accurate estimates for δ_d, f_{off}, and t_s and can achieve a good BER within 10 to 20 observation symbols. To obtain the BER close to the lower bound, the optimum estimator requires 10 observation symbols, whereas 40 observation symbols are still insufficient for the suboptimum estimator.

6.5 Conclusions

An OFDM signal has cyclostationary property. Making effective use of this property, we can nicely carry out the channel parameter estimation essential for demodulation of transmitted data. This chapter presented a blind maximum likelihood-based joint DFT window timing/frequency offset/DFT window width estimation method. The numerical results have shown the superiority of the proposed method.

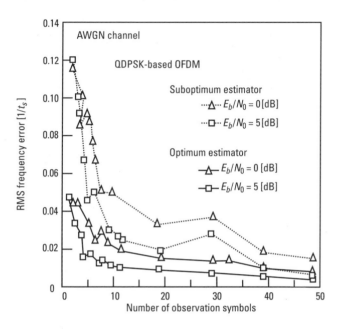

Figure 6.5 RMS frequency error in an AWGN channel.

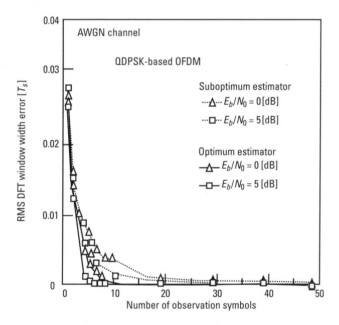

Figure 6.6 RMS DFT window width error in an AWGN channel.

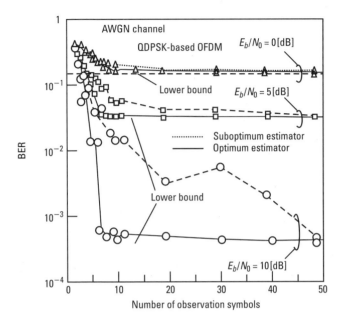

Figure 6.7   BER in an AWGN channel.

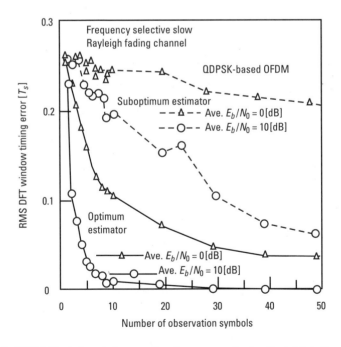

Figure 6.8   RMS DFT window timing error in a frequency selective Rayleigh fading channel.

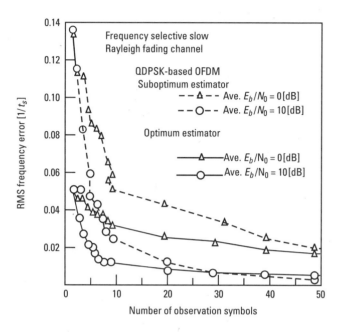

Figure 6.9   RMS frequency error in a frequency selective Rayleigh fading channel.

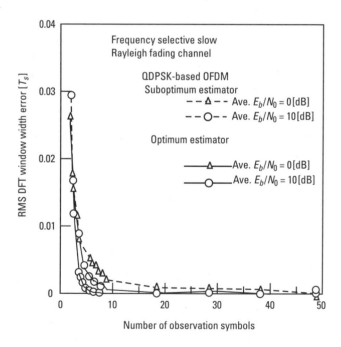

Figure 6.10   RMS DFT window width error in a frequency selective Rayleigh fading channel.

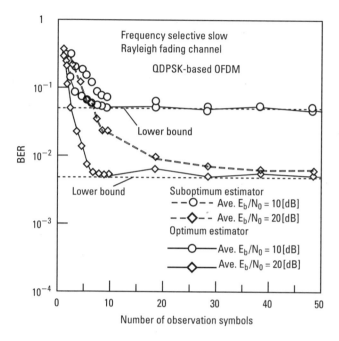

Figure 6.11   BER in a frequency selective Rayleigh fading channel.

References

[1]   Mouri, M., et al., "Joint Symbol-Timing and Frequency Offset Estimation Scheme for Multicarrier Modulation System (in Japanese)," *IEICE Technical Report*, RCS95-70, 1995, pp. 9–16.

[2]   Daffara, F., and O. Adami, "A New Frequency Detector for Orthogonal Multicarrier Transmission Techniques," *Proc. IEEE VTC'95*, 1995, pp. 804–809.

[3]   Gardner, W. A., *Cyclostationarity in Communications and Signal Processing*, New York: IEEE Press, 1994.

[4]   Okada, M., et al., "A Maximum Likelihood Symbol Timing, Symbol Period, and Frequency Offset Estimator for Orthogonal Multicarrier Modulation Signals," *Proc. IEEE ICT'96*, 1996, pp. 596–601.

[5]   Okada, M., et al., "Optimum Synchronization of Orthogonal Multicarrier Modulated Signals," *Proc. IEEE PIMRC'96*, 1996, pp. 863–867.

7

Coded OFDM Scheme to Gain Frequency Diversity Effect

7.1 Introduction

In Chapters 4–6, we discussed the BER performance of several OFDM systems in frequency selective fading channels, where we did not take into consideration any channel coding schemes. In this sense, we can call them "uncoded OFDM systems." Frequency selective fading gives a distortion to the channel frequency response, so when a signal is sent through a frequency selective fading channel, some subcarriers experience high attenuation whereas others experience low attenuation. Although there are indeed subcarriers with low attenuation and no errors, the *average* BER of an *uncoded* OFDM system is determined by the worse BERs in subcarriers with high attenuation, so its BER performance is very close to that of an *uncoded* single carrier system in a frequency nonselective fading channel. We can imagine that if we employ a channel coding scheme over subcarriers, we can improve the BER by means of a frequency diversity effect.

In this chapter, we discuss "the channel coding effect" on the BER performance of an OFDM system in frequency selective fading channels. First, Section 7.2 outlines a convolutional encoding/Viterbi decoding scheme. After explaining the role of interleaving in fading channels, Sections 7.3 and 7.4 show the principles of symbol interleaving and bit interleaving schemes in detail, with much emphasis on how to calculate the path metric in each scheme. Then, Section 7.5 discusses the BER of the interleaved coded OFDM schemes and compares the effect of symbol interleaving with that of bit interleaving. Finally, Section 7.6 draws conclusions.

7.2 Convolutional Encoding/Viterbi Decoding

A convolutional encoding is done by passing the information sequence to be transmitted through a linear finite-state shift register [1, 2]. The shift register is composed of K (k-bit) stages and n linear algebraic function generators. In this case, the input data is shifted into and along the shift register k bits at a time, and the number of output bits for each k-bit input sequence is n bits. Therefore, the code rate, which is an important parameter to describe the convolutional code, is defined as $R_c = k/n$. Another important parameter is the constraint length of the convolutional code, which is defined as $K + 1$.

Figure 7.1 shows a convolutional encoder with $K = 6$, $k = 1$, and $n = 2$, so $R_c = 1/2$ [the outputs $b_{i1} b_{i2}$ for the input a_i ($i = 1, 2, \ldots$)], which we will use in this chapter. This encoder is characterized by the following two generators:

$$g_1 = [1011011] \tag{7.1}$$

$$g_2 = [1111001] \tag{7.2}$$

where the position of "1" in (7.1) and (7.2) shows the position of a stage that has a connection to the upper and lower function generators (adders), respectively. This convolutional code also has the minimum free distance $d_{free} = 10$, which largely determines the attainable BER performance.

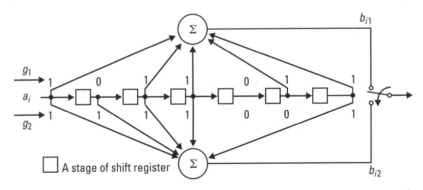

Figure 7.1 A convolutional encoder with $K = 6$, $k = 1$, and $n = 2$.

Now, define the convolutional encoder output vector, the transmitted signal vector, and the received signal vector as

$$\mathbf{b}_i = (b_{i1}, \, b_{i2}) \qquad (7.3)$$

$$\mathbf{c}_i = (c_{i1}, \, c_{i2}) \qquad (7.4)$$

$$\mathbf{r}_i = (r_{i1}, \, r_{i2}) \qquad (7.5)$$

The vectors $\mathbf{b}_i$ and $\mathbf{c}_i$ are connected with a one-to-one labeling map μ as [3]:

$$\mathbf{c}_i = \mu(\mathbf{b}_i) \qquad (7.6)$$

for instance,

$$\mathbf{c}_i = \begin{cases} (1, \, 1), & \mathbf{b}_i = (0, \, 0) \\ (-1, \, 1), & \mathbf{b}_i = (0, \, 1) \\ (-1, \, -1), & \mathbf{b}_i = (1, \, 1) \\ (1, \, -1), & \mathbf{b}_i = (1, \, 0) \end{cases} \qquad (7.7)$$

The Viterbi algorithm is an efficient sequential search algorithm that performs the following maximum likelihood sequence detection:

$$\underline{\hat{\mathbf{b}}} = \arg \max_{\underline{\mathbf{b}} \in \mathbf{B}} \sum_i \log p_{\mathbf{z}_k}\big(\mathbf{r}_i | \mathbf{c}_i = \mu(\mathbf{b}_i)\big) \qquad (7.8)$$

where the underline of * is a sequence of *, $\mathbf{B}$ is a set that is composed of all possible encoder output vector sequences, and $p_{\mathbf{z}_i}\big(\mathbf{r}_i | \mu(\mathbf{b}_i)\big)$ is the transition p.d.f. with a vector parameter $\mathbf{z}_i$.

Figure 7.2 shows a trellis diagram of the Viterbi decoding for the convolutional code in Figure 7.1, where $S_0 \ldots S_{63}$ show states. The number of states is given by 2^K.

7.3 Symbol Interleaved Coded OFDM Scheme

Convolutional codes are generally designed for channels where the errors are random. However, in a multipath fading channel, signal fading causes

Received signal vector

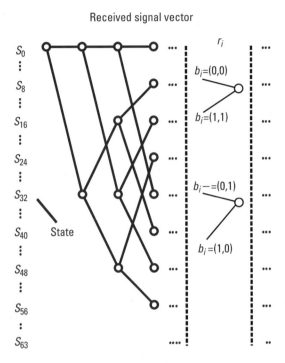

Figure 7.2 A trellis diagram for Viterbi decoding.

bursty errors, because the time variant nature of the channel makes the level of the received signal fall below the noise level and the fade duration contains several transmitted data. Interleaving/deinterleaving is an effective method that transforms the bursty channel into a channel with random errors.

Figure 7.3 shows a method of interleaving/deinterleaving. At the transmitter, the encoder outputs are reordered by the interleaver with operations of "write-in" and "read-out" and then transmitted over the bursty channel. At the receiver, the received data is again reordered by the deinterleaver in proper order and then passed to the Viterbi decoder. In this way, interleaving/deinterleaving spreads out bursty errors to make errors within a code word be random.

Figure 7.4 shows the block diagram of a symbol interleaved coded OFDM scheme, where the symbol interleaver performs symbol-wise reordering. Assume that the ith encoder output $\mathbf{b}_i$ is transmitted in the form of the i'th symbol $\mathbf{c}_{i'}$ over the k'th subcarrier, and that the received envelope of the k'th subcarrier is $\mathbf{z}_{k'} = (z_{k'1}, z_{k'2})$. Note that in the symbol interleaving, the two information bits in $\mathbf{b}_i$ are transmitted over the same subcarrier.

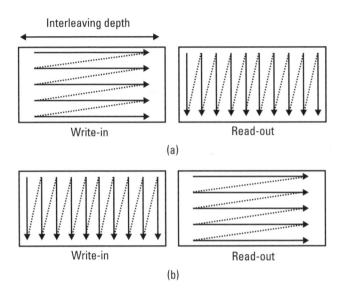

Figure 7.3 (a) Interleaving; and (b) deinterleaving.

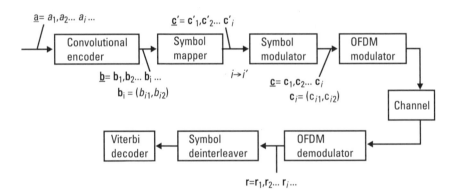

Figure 7.4 Block diagram of a symbol interleaved coded OFDM scheme.

Newly defining a labeling map between $\mathbf{b}_i$ and $\mathbf{c}_{i'}$ as

$$\mathbf{c}_i = \mu(\mathbf{b}_i) \qquad (7.9)$$

the maximum likelihood decoding is made according to

$$\hat{\underline{\mathbf{b}}} = \arg \max_{\underline{\mathbf{b}} \in \mathbf{B}} \sum_i \log p_{\mathbf{z}_{k'}}\left(\mathbf{r}_{i'} \,|\, \mathbf{c}_{i'} = \mu(\mathbf{b}_i)\right) \qquad (7.10)$$

The transition p.d.f. is written as

$$p_{z_{k'}}\left(\mathbf{r}_{i'}|\mathbf{c}_{i'} = \mu(\mathbf{b}_i)\right) = \left(\frac{1}{\sqrt{2\pi}\sigma_n}\right)^2 \exp\left(-\frac{|\mathbf{r}_{i'} - \mathbf{z}_{k'}\mathbf{c}_{i'}|^2}{2\sigma_n^2}\right)$$

(7.11)

where σ_n^2 is the noise power and $\mathbf{z}_{k'}\mathbf{c}_{i'}$ is defined as

$$\mathbf{z}_{k'}\mathbf{c}_{i'} = \left(z_{k'1}c_{i'1} - z_{k'2}c_{i'2}, \; z_{k'1}c_{i'2} + z_{k'2}c_{i'1}\right)$$

(7.12)

From (7.11), we can see

$$\log p_{z_{k'}}\left(\mathbf{r}_{i'}|\mu(\mathbf{b}_i)\right) \propto -\left|\mathbf{r}_{i'} - \mathbf{z}_{k'}\mathbf{c}_{i'}\right|^2$$

(7.13)

therefore, (7.10) is finally simplified to

$$\hat{\underline{\mathbf{b}}} = \arg\min_{\underline{\mathbf{b}} \in \mathbf{B}} \sum_i \left|\mathbf{r}_{i'} - \mathbf{z}_{k'}\mathbf{c}_{i'}\right|^2$$

(7.14)

Here, define the branch metric for $\mathbf{b}_i = (b_{i1}, b_{i2})$, which corresponds to the Euclidean distance between $\mathbf{r}_{i'}$ and $\mathbf{z}_{k'}\mathbf{c}_{i'}$, as

$$d_{b_{i1}b_{i2}} = \left|\mathbf{r}_{i'} - \mathbf{z}_{k'}\mu((b_{i1}, b_{i2}))\right|$$

(7.15)

Figure 7.5 shows a received QPSK signal constellation at the k'th subcarrier. The Viterbi decoding algorithm, based on the maximum likelihood criterion, can effectively select a path (an information bit sequence) with the smallest path metric, by selecting a most probable path with the smallest branch metric, namely, the smallest Euclidean distance.

7.4 Bit Interleaved Coded OFDM Scheme

Figure 7.6 shows the block diagram of a bit interleaved coded OFDM scheme, where the bit interleaver performs bitwise reordering. Note that, in this case, the two information bits in $\mathbf{b}_i$, in the forms of parts of two symbols, are transmitted over different subcarriers. Therefore, we need to define two new labeling maps.

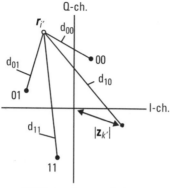

QPSK signal constellation
at the k'th subcarrier

Figure 7.5 Calculation of Euclidean distance in the symbol interleaved coded OFDM scheme.

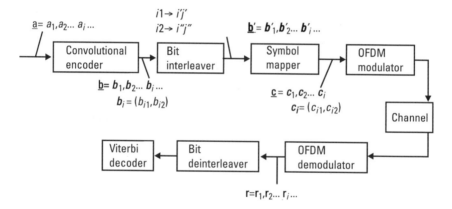

Figure 7.6 Block diagram of a bit interleaved coded OFDM scheme.

Assuming that b_{i1} is mapped to the first bit of the i'th symbol and b_{i2} to the second bit of the i''th symbol as

$$\mathbf{c}_{i'} = (\mu_1(b_{i1}),\, c_{i'2}) \tag{7.16}$$

$$\mathbf{c}_{i''} = (c_{i''1},\, \mu_2(b_{i2})) \tag{7.17}$$

the maximum likelihood decoding is made according to

$$\hat{\underline{b}} = \arg\max_{\underline{b} \in \mathbf{B}} \sum_i \log p_{\mathbf{z}_{k'},\,\mathbf{z}_{k''}}(\mathbf{r}_{i'},\, \mathbf{r}_{i''}|\mu_1(b_{i1}),\, \mu_2(b_{i2})) \tag{7.18}$$

The transition p.d.f. is written as

$$p_{z_{k'},z_{k''}}\left(\mathbf{r}_{i'},\ \mathbf{r}_{i''}|\mu_1(b_{i1}),\ \mu_2(b_{i2})\right) =$$
$$\left\{p(c_{i'2}=1)p_{z_{k'}}\left(\mathbf{r}_{i'}|\mathbf{c}_{i'}=(\mu_1(b_{i1}),\ 1)\right)\right.$$
$$\left.+\ p(c_{i'2}=-1)p_{z_{k'}}\left(\mathbf{r}_{i'}|\mathbf{c}_{i'}=(\mu_1(b_{i1}),\ -1)\right)\right\} \qquad (7.19)$$
$$\times\left\{p(c_{i''1}=1)p_{z_{k''}}\left(\mathbf{r}_{i''}|\mathbf{c}_{i''}=(1,\ \mu_2(b_{i2}))\right)\right.$$
$$\left.+\ p(c_{i''1}=-1)p_{z_{k''}}\left(\mathbf{r}_{i''}|\mathbf{c}_{i''}=(-1,\ \mu_2(b_{i2}))\right)\right\}$$

For a uniform input distribution,

$$p(c_{i'2}=1)=p(c_{i'2}=-1)=1/2 \qquad (7.20)$$

$$p(c_{i''1}=1)=p(c_{i''1}=-1)=1/2 \qquad (7.21)$$

and when SNR is high, we can approximate (7.19) as [3]

$$p_{z_{k'},z_{k''}}\left(\mathbf{r}_{i'},\ \mathbf{r}_{i''}|\mu_1(b_{i1}),\ \mu_2(b_{i2})\right) \approx$$
$$\left\{\max\left[p_{z_{k'}}\left(\mathbf{r}_{i'}|\mathbf{c}_{i'}=(\mu_1(b_{i1}),\ 1)\right),\right.\right.$$
$$\left.\left.p_{z_{k'}}\left(\mathbf{r}_{i'}|\mathbf{c}_{i'}=(\mu_1(b_{i1}),\ -1)\right)\right]/2\right\} \qquad (7.22)$$
$$\times\left\{\max\left[p_{z_{k''}}\left(\mathbf{r}_{i''}|\mathbf{c}_{i''}=(1,\ \mu_2(b_{i2}))\right),\right.\right.$$
$$\left.\left.+\ p_{z_{k''}}\left(\mathbf{r}_{i''}|\mathbf{c}_{i''}=(-1,\ \mu_2(b_{i2}))\right)\right]/2\right\}$$

Therefore, (7.18) can be finally simplified to

$$\hat{\underline{b}} = \arg\min_{\underline{b}\in\mathbf{B}}\sum_i\left\{\min\left[\left|\mathbf{r}_{i'}-\mathbf{z}_{k'}\mathbf{c}_{i'}\right|^2_{c_{i'2}=1},\ \left|\mathbf{r}_{i'}-\mathbf{z}_{k'}\mathbf{c}_{i'}\right|^2_{c_{i'2}=-1}\right]\right. \qquad (7.23)$$
$$\left.+\ \min\left[\left|\mathbf{r}_{i''}-\mathbf{z}_{k''}\mathbf{c}_{i''}\right|^2_{c_{i''1}=1},\ \left|\mathbf{r}_{i''}-\mathbf{z}_{k''}\mathbf{c}_{i''}\right|^2_{c_{i''1}=-1}\right]\right\}$$

Here, define the branch metric for $\mathbf{b}_i=(b_{i1},\ b_{i2})$ as

$$d_{b_{i1}b_{i2}}=d_{b_{i1}*}+d_{*b_{i2}} \qquad (7.24)$$

$$d_{b_{i1}}* = \min\left[\,|\mathbf{r}_{i'} - \mathbf{z}_{k'}(\mu_1(b_{i1},\,1))|,\,|\mathbf{r}_{i'} - \mathbf{z}_{k'}(\mu_1(b_{i1},\,-1))|\,\right]$$
(7.25)

$$d*_{b_{i2}} = \min\left[\,|\mathbf{r}_{i''} - \mathbf{z}_{k''}(1,\,\mu_2(b_{i2}))|,\,|\mathbf{r}_{i''} - \mathbf{z}_{k''}(-1,\,\mu_2(b_{i2}))|\,\right]$$
(7.26)

Figure 7.7 shows the relation between the branch metric and two received QPSK signal constellations at the k'th and k''th subcarriers. The way of calculating the branch metric can be easily extended to a more general modulation scheme, such as M-ary QAM and so on [3].

7.5 Numerical Results and Discussions

We assume an OFDM system adopted in high-rate wireless LAN standards [4, 5]. Figure 7.8 shows the subcarrier arrangement. The subcarriers from 0 to 63 are generated by the 64-point inverse fast Fourier transform (IFFT). Among those, the 12 subcarriers from 0 to 5, 32, and 59 to 63 are called "virtual subcarriers," which are not used for actual data transmission (in other words, the virtual subcarriers transmit 0). The four subcarriers of 11, 25, 39, and 53 are "pilot subcarriers," which always transmit known symbols to adjust the frequency of the local oscillator at the receiver, and the remaining 48 subcarriers are "data subcarriers." After the 48 data subcarriers are

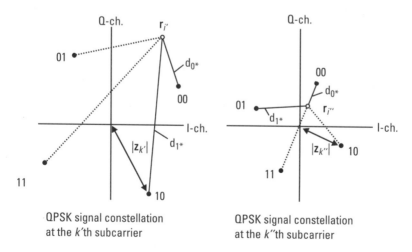

QPSK signal constellation
at the k'th subcarrier

QPSK signal constellation
at the k''th subcarrier

Figure 7.7 Calculation of Euclidean distance in the symbol interleaved coded OFDM scheme.

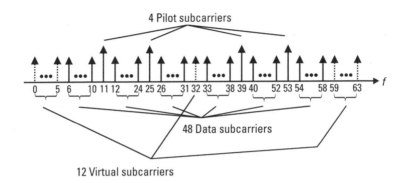

Figure 7.8 Subcarrier arrangement of an OFDM scheme.

generated by means of 64-point IFFT, the 16-sample long guard interval is added to the generated waveform (also see Figure 5.26). Interleaving is done within one OFDM symbol. Table 7.1 shows the transmission parameters to demonstrate the BER performance.

Figure 7.9 shows the BER in a frequency nonselective Rayleigh fading channel, namely, where there is one path in the multipath delay profile. The theoretical BER is given by [2]

$$P_{b,fading}^{Q,coherent} = \sum_{d=d_{free}}^{\infty} \beta_d P(d) \qquad (7.27)$$

$$P(d) = \int_0^{\infty} P(\gamma_b')^d \sum_{k=0}^{d-1} \binom{d-1+k}{k} (1 - P(\gamma_b'))^k p(\gamma_b') \, d\gamma_b'$$

$$(7.28)$$

Table 7.1
Transmission Parameters for BER Evaluation

Number of data subcarriers	48 (64-point IFFT)
Guard interval length	16 [samples]
Modulation/demodulation	CQPSK
Data burst length	10 [OFDM symbols]
Subcarrier recovery	Perfect
Channel model	Frequency nonselective Rayleigh fading, 2-path and 4-path i.i.d. frequency selective Rayleigh fading (Delay of each path is uniformly distributed within the guard interval.)

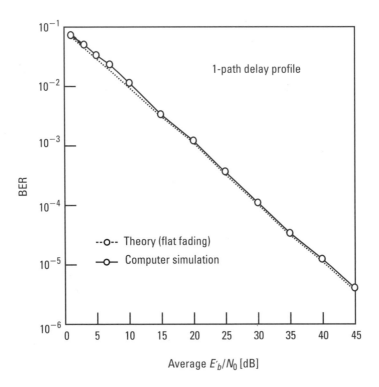

Figure 7.9 BER in a frequency nonselective Rayleigh fading channel.

where $P(\gamma_b')$ is given by (4.11), namely, the BER of coherent QPSK when γ_b' is given, and $p(\gamma_b')$ is the p.d.f. of γ_b', which is given by (4.18). In addition, $\{\beta_d\}$ are the weighting coefficients calculated from the transfer function of the convolutional code. Table 7.2 shows the values of $\{\beta_d\}$ for the convolutional code with the generators given by (7.1) and (7.2) [6]. In the calculation of (7.27), the summation was upper limited by $d = d_{free} + 4 = 14$.

Table 7.2
$\{\beta_d\}$ for the Convolutional Code Given by (7.1) and (7.2)

$d = d_{free} \ (= 10)$	36
$d = d_{free} + 1$	0
$d = d_{free} + 2$	211
$d = d_{free} + 3$	0
$d = d_{free} + 4$	1,404

The computer simulation result agrees well with the theoretical one. For the frequency nonselective fading channel where all the subcarriers are subject to the same attenuation at a time, there is no diversity effect even if we employ channel coding. This is very clear from Figure 7.9, where the BER reduces by factor 10^{-1} when the average E_b'/N_0 gains +10 dB.

Figures 7.10 and 7.11 show the BER in a 3-path i.i.d. frequency selective Rayleigh fading channel for symbol interleaving and bit interleaving, respectively. The theoretical lower bound is given by the BER expression for the Lth order diversity (with $L = 3$) [2]:

$$P_{b,\,fading}^{Q,\,coherent} = \left(\frac{1}{4\overline{\gamma_b'}}\right)^L \left(\begin{array}{c} 2L - 1 \\ L \end{array}\right) \qquad (7.29)$$

Figures 7.12 and 7.13 show the BER against the interleaving depth for symbol interleaving and bit interleaving, respectively. From Figures 7.10

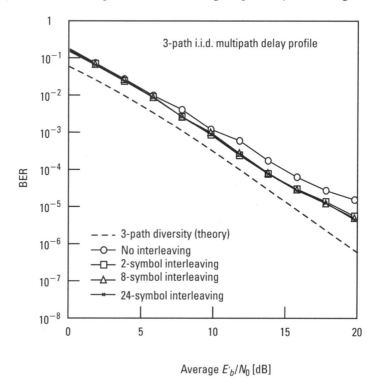

Figure 7.10 BER of symbol interleaved coded OFDM scheme in a 3-path i.i.d. frequency selective Rayleigh fading channel.

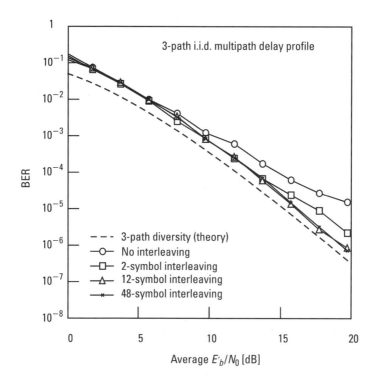

Figure 7.11 BER of bit interleaved coded OFDM scheme in a 3-path i.i.d. frequency selective Rayleigh fading channel.

and 7.12, we can see that the interleaving depth of four symbols ($\times$ 12 symbols) is enough to obtain good BER performance. Even when we set the interleaving depth to more than four symbols, there is no significant performance gain obtained. On the other hand, from Figures 7.11 and 7.13, we can see that the interleaving depth of 8 bits ($\times$ 12 bits) is enough to obtain good BER performance.

Figure 7.14 compares the BER between a symbol interleaving depth of four symbols and a bit interleaving depth of 8 bits. The performance with 8-bit interleaving is superior to one with four-symbol interleaving. This is because in symbol interleaving, the upper bit and lower bit in one encoded output are transmitted over the same subcarrier, so pairwise errors tend to occur, whereas for bit interleaving, they are transmitted over different subcarriers, so pairwise errors do not occur.

Figures 7.15 and 7.16 show the BER in a 4-path i.i.d. frequency selective Rayleigh fading channel for symbol interleaving and bit interleaving, respectively. The theoretical lower bound is given by (7.29) with $L = 4$.

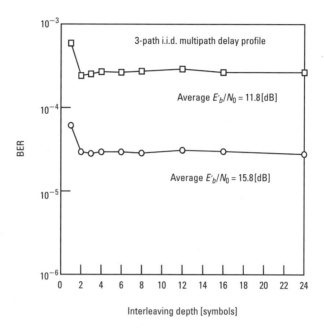

Figure 7.12 BER against interleaving depth for symbol interleaved coded OFDM scheme.

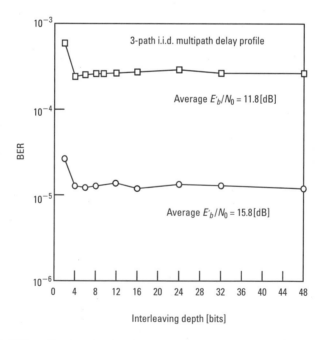

Figure 7.13 BER against interleaving depth for bit interleaved coded OFDM scheme.

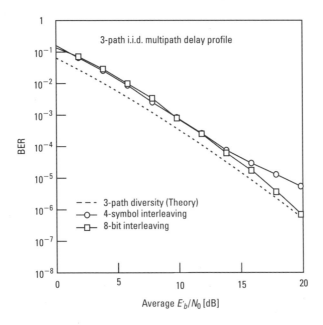

Figure 7.14 BER comparison between symbol and bit interleaved coded OFDM schemes.

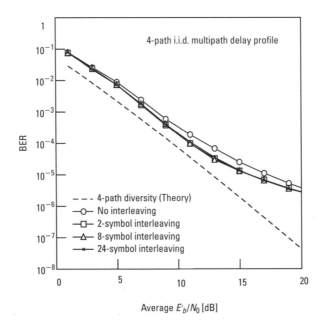

Figure 7.15 BER of symbol interleaved coded OFDM scheme in a 4-path i.i.d. frequency selective Rayleigh fading channel.

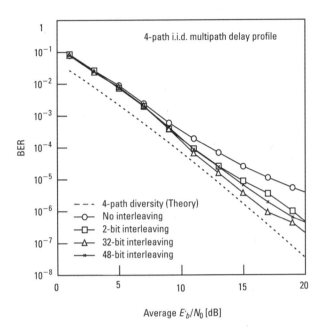

Figure 7.16 BER of bit interleaved coded OFDM scheme in a 4-path i.i.d. frequency selective Rayleigh fading channel.

Furthermore, Figures 7.17 and 7.18 show the BER against the interleaving depth for symbol interleaving and bit interleaving, respectively. From Figures 7.15 and 7.17, we can see that, even if we increase the symbol interleaving depth, we cannot much improve the BER performance. This may be because of pairwise errors. On the other hand, from Figures 7.16 and 7.18, we can see that the BER of the bit interleaved coded OFDM scheme with an appropriate bit interleaving depth is close to the lower bound, and that the BER is sensitive to the bit interleaving depth chosen and there is an optimum interleaving depth to minimize the BER for the given channel parameter setting.

Finally, Figure 7.19 compares the BER between a symbol interleaving depth of four symbols and a bit interleaving depth of 8 bits. The performance with 8-bit interleaving is much superior to that with 4-symbol interleaving.

7.6 Conclusions

This chapter discussed a coded OFDM scheme to gain a frequency diversity effect in frequency selective fading channels. A convolutional coding/Viterbi

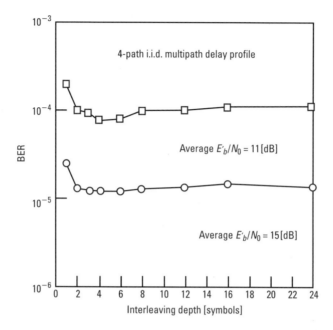

Figure 7.17 BER against interleaving depth for symbol interleaved coded OFDM scheme.

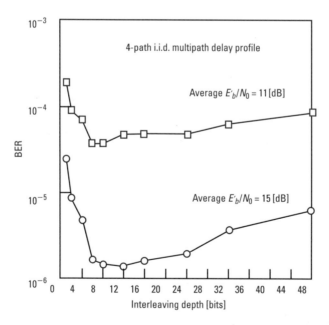

Figure 7.18 BER against interleaving depth for bit interleaved coded OFDM scheme.

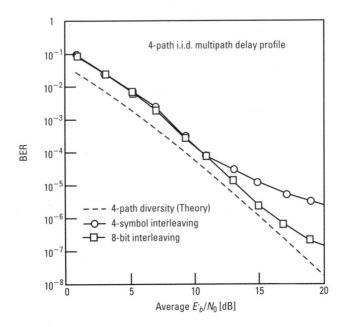

Figure 7.19 BER comparison between symbol and bit interleaved coded OFDM schemes.

decoding has been assumed, and the effect of symbol interleaving and bit interleaving has been fully examined for randomizing bursty errors induced in the fading channels. For a given system specification and channel condition, a bit interleaved convolutional coded OFDM scheme, with an appropriate interleaving depth chosen, can achieve good BER performance even in such severe channels.

References

[1]   Clark, Jr., G. C., and J. B. Cain, *Error-Correcting Coding for Digital Communications,* New York: Plenum Press, 1981.

[2]   Proakis, J. G., *Digital Communications, Fourth Edition,* New York: McGraw-Hill, 2001.

[3]   Caire, G., G. Taricco, and E. Biglieri, "Bit-Interleaved Coded Modulation," *IEEE Trans. Inform. Theory.,* Vol. IT-44, No. 3, May 1998, pp. 927–946.

[4]   van Nee, R., and R. Prasad, *OFDM for Wireless Multimedia Communications,* Norwood, MA: Artech House, 2000.

[5]   van Nee, R., et al., "New High-Rate Wireless LAN Standards," *IEEE Commun. Mag.,* Vol. 37, No. 12, December 1999, pp. 82–88.

[6]   Conan, J., "The Weight Spectra of Some Short Low-Rate Convolutional Codes," *IEEE Trans. Commun.,* Vol. COM-32, No. 9, September 1984, pp. 1050–1053.

8

Applications of OFDM

8.1 Introduction

The previous chapters gave all the materials required for understanding current OFDM-based systems. Chapter 2 gave the prerequisite knowledge on radio channels, Chapter 3 the principle and history of OFDM, Chapter 4 the OFDM characteristics, including the theoretical analysis in frequency selective fading channels, Chapter 5 and 6 the synchronization issues, and Chapter 7 the forward error correction (FEC) to gain a frequency diversity effect.

This chapter describes several applications of OFDM scheme in various systems. Section 8.2 introduces the application in digital broadcasting, such as DAB, DVB-T, and ISDB-T. Section 8.3 describes the application in 5 GHz-band wireless LANs, such as IEEE 802.11a, HIPERLAN type 2, and MMAC. Section 8.4 introduces the applications in wireless LANs that are under development for standardization by the end of 2002, such as IEEE 802.11g, IEEE 802.11h, and IEEE 802.16a.

8.2 Digital Broadcasting

8.2.1 Digital Audio Broadcasting

Advances in high fidelity (hi-fi) digital recording techniques triggered a digital revolution in the late 1980s in sound broadcasting technology. The hi-fi sound sources were not only suitable for indoor but also outdoor use, and they gave rise to the requirement for mobile reception of digital audio signals. This requirement, which was considered impossible in the early 1980s because of the multipath problem, was suddenly fulfilled by the use of OFDM.

Digital audio broadcasting (DAB) was specified between 1988 and 1992, with its introduction in Europe scheduled for the late 1990s. Many DAB field trials were carried out by broadcasters in Europe, including DAB single frequency operations in Munich, Germany, DAB test operations in video mode using the telecommunication satellite KOPERNIKUS, field trials in L-band, and so on. In addition to these, a lot of DAB measurements were also carried out, on electromagnetic field strength, channel impulse response, invulnerability against cochannel interference, bit error rate, and so on. Table 8.1 shows the three modes defined in EUREKA 147 DAB [1].

8.2.2 Terrestrial Digital Video Broadcasting

In Europe, based on the successful results from the DAB field trials and measurements, terrestrial digital video broadcasting (DVB-T), with use of OFDM, was standardized by the European Telecommunications Standards Institute (ETSI) in 1996. Table 8.2 shows the two modes defined in the DVB-T [2]. In 1998, the DVB-T was first adopted in the United Kingdom, with multifrequency network (MFN) use, 2k mode, 64 QAM, 7-μs guard interval, $R_c = 2/3$-convolutional code, and 24.13-Mbps information transmission rate.

Table 8.1
DAB Parameters

Parameter Mode	Mode 1	Mode 2	Mode 3
Bandwidth	1.536 MHz	1.536 MHz	1.536 MHz
Number of subcarriers	1,546	768	384
Modulation		DQPSK	
Useful symbol length (t_s)	1 ms	250 μs	125 μs
Subcarrier separation (Δf)	3.968 kHz	1.984 kHz	0.992 kHz
Guard interval length (Δ_G)	$t_s/4$ (250 μs)	$t_s/4$ (62.5 μs)	$t_s/4$ (31.25 μs)
FEC	Convolutional code		
Information transmission rate		2.4 Mbps	

Table 8.2
DVB-T Parameters

Parameter Mode	2k		8k
Bandwidth	7.61 MHz		7.61 MHz
Number of subcarriers	1,705		6,817
Modulation	QPSK	16 QAM	64 QAM
Useful symbol length (t_s)	224 μs		896 μs
Subcarrier separation (Δf)	4.464 kHz		1.116 kHz
Guard interval length (Δ_G)	$t_s/4$ $t_s/8$ $t_s/16$ $t_s/32$		$t_s/4$ $t_s/8$ $t_s/16$ $t_s/32$
	56 μs 28 μs 14 μs 7 μs		224 μs 112 μs 56 μs 28 μs
FEC (inner code)	Convolutional code (R = 1/2, 2/3, 3/4, 5/6, 7/8)		
FEC (outer code)	Reed-Solomon code (204, 188)		
Interleaving	Time-frequency domain bit interleaving		
Information transmission rate	4.98–31.67 Mbps		
Required C/N	3.1 dB–20.1 dB		

8.2.3 Terrestrial Integrated Services Digital Broadcasting

In Japan, the Association of Radio Industries and Businesses (ARIB) standardized terrestrial integrated services digital broadcasting (ISDB-T) in June 2000. Table 8.3 shows the three modes defined in the ISDB-T for television and Table 8.4 for audio [3]. The commercial service-in is scheduled in 2003.

It is very interesting to compare Table 8.2 and Table 8.3. The number of subcarriers in the DVB-T is a bit more than that in the ISDB-T with wider occupied bandwidth, and the DVB-T uses only coherent demodulation schemes, whereas the ISDB-T uses not only coherent but also differential demodulation schemes. The narrower occupied bandwidth and the use of differential detection in the ISDB-T are both designed for mobile reception, because they can give it robustness against frequency selective fading even with low SNR.

In addition to these, the ISDB-T uses an interesting pilot symbol insertion method to support mobile reception. Figure 8.1 shows the time-frequency structure of an ISDB-T-based OFDM signal. In the pilot symbol layout, there are two kinds of pilot symbol patterns, such as the continuous pilot pattern on the subcarrier with the highest frequency and the scattered pilot pattern where pilot symbols are inserted at certain subcarriers and certain time samples. The channel transfer function essential for coherent demodulation is estimated from interpolation of the scattered pilot symbols along the frequency axis, and the scattered pilot symbols are transmitted at

Table 8.3
ISDB-T Parameters (Television)

Parameter Mode	Mode 1	Mode 2	Mode 3
Bandwidth	5.575 MHz	5.573 MHz	5.572 MHz
Number of subcarriers	1,405	2,809	5,617
Modulation	QPSK 16 QAM 64 QAM DQPSK		
Useful symbol length (t_S)	252 μs	504 μs	1,008 μs
Subcarrier separation (Δf)	3.968 kHz	1.984 kHz	0.992 kHz
Guard interval length (Δ_G)	$t_S/4$ (63 μs)	$t_S/4$ (126 μs)	$t_S/4$ (252 μs)
	$t_S/8$ (31.5 μs)	$t_S/8$ (63 μs)	$t_S/8$ (126 μs)
	$t_S/16$ (15.75 μs)	$t_S/16$ (31.5 μs)	$t_S/16$ (63 μs)
	$t_S/32$ (7.875 μs)	$t_S/32$ (15.75 μs)	$t_S/32$ (31.5 μs)
FEC (inner code)	Convolutional code ($R = $ 1/2, 2/3, 3/4, 5/6, 7/8)		
FEC (outer code)	Reed-Solomon Code (204, 188)		
Interleaving	Time-frequency domain bit interleaving		
Information transmission rate	3.65–23.2 Mbps		
Required C/N	3.1 dB–20.1 dB		

every time sample, so they can easily track the time variation of the channel transfer function.

8.3 5 GHz-Band Wireless LANs

In 1998, the IEEE 802.11 standardization group decided to select OFDM as a basis for its new 5-GHz wireless LAN standard, which supports data transmission rates from 6 to 56 Mbps. In the DVB-T and ISDB-T, which are mentioned in Section 8.2, OFDM is used in continuous transmission mode for the purpose of broadcasting. This new standard, called "IEEE 802.11a," is the first to use OFDM in packet transmission mode [4].

Following the IEEE 802.11 decision, ETSI adopted OFDM in the standard of HIPERLAN/2 [5], as well as ARIB in the standard of MMAC [6, 7]. Since then, the three bodies have worked in close cooperation to ensure that differences between the three standards are kept to a minimum, enabling the manufacturing of equipment that can be used worldwide.

The main difference between IEEE 802.11a and HIPERLAN type 2 is in the medium access control (MAC). The IEEE 802.11a uses a distributed MAC based on carrier sense multiple access with collision avoidance (CSMA/CA), whereas the HIPERLAN type 2 uses a centralized and scheduled MAC based on time division multiple access with dynamic slot assignment

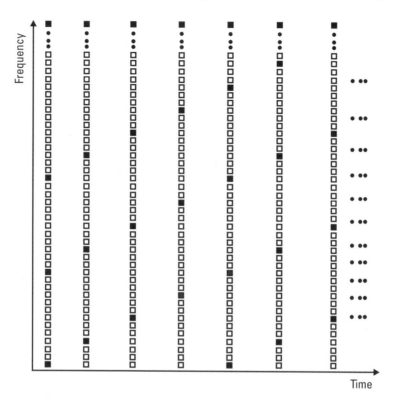

Time

Figure 8.1 Time-frequency structure of an ISDB-T-based OFDM signal.

(TDMA/DSA). The MMAC supports both of these MACs. In terms of the physical layer (PHY), there are only a few minor differences among the three standards. Table 8.5 shows the system parameters for the IEEE 802.11a [4] and the MMAC [6] and Table 8.6 for the HIPERLAN type 2 [5] and the MMAC [7].

Figure 8.2 shows the time-frequency structure of an IEEE 802.11a-based OFDM packet [8, 9]. To achieve packet mode transmission, a preamble is attached at the head of the payload. The preamble is composed of patterns A, B, and C, which are all known training signals. Usually, pattern A (80 sample-long) is used for automatic gain control, pattern B (80 sample-long) for FFT timing synchronization and coarse frequency offset compensation, and pattern C (160 sample-long) for subcarrier recovery to carry out coherent demodulation. In patterns A and B, there are pilot symbols inserted at almost every four subcarriers, which is an extension of Schmidl's method (see Section 5.2).

Table 8.4
ISDB-T Parameters (Audio)

Bandwidth	429 kHz*		
	1.27 MHz**		
Number of subcarriers	109*	217*	433*
	325**	649**	1,297**
Modulation	QPSK 16 QAM 64 QAM DQPSK		
Useful symbol length (t_S)	252 μs	504 μs	1,008 μs
Subcarrier separation (Δf)	3.968 kHz	1.984 kHz	0.992 kHz
Guard interval length (Δ_G)	$t_S/4$ (63 μs)	$t_S/4$ (126 μs)	$t_S/4$ (252 μs)
	$t_S/8$ (31.5 μs)	$t_S/8$ (63 μs)	$t_S/8$ (126 μs)
	$t_S/16$ (15.75 μs)	$t_S/16$ (31.5 μs)	$t_S/16$ (63 μs)
	$t_S/32$ (7.875 μs)	$t_S/32$ (15.75 μs)	$t_S/32$ (31.5 μs)
FEC (inner code)	Convolutional code (R = 1/2, 2/3, 3/4, 5/6, 7/8)		
FEC (outer code)	Reed-Solomon Code (204, 188)		
Interleaving	Time-frequency domain bit interleaving		
Information transmission rate	280.8–840 Kbps		

(* 1-segment transmission, ** 3-segment transmission)

Table 8.5
IEEE 802.11a and MMAC Parameters

Channel spacing	20 MHz
Bandwidth	16.56 MHz (−3 dB)
Number of subcarriers	52
Number of pilot subcarriers	4
Useful symbol length (t_S)	3.2 μs
Subcarrier separation (Δf)	312.5 kHz
Guard interval length (Δ_G)	800 ns
FEC	Convolutional code
Interleaving	Frequency domain bit interleaving (within one OFDM symbol)
Information transmission rate/ Modulation/coding rate	6 Mbps (BSPK, R_c = 1/2)
	9 Mbps (BSPK, R_c = 3/4)
	12 Mbps (QSPK, R_c = 1/2)
	18 Mbps (QSPK, R_c = 3/4)
	24 Mbps (16 QAM, R_c = 1/2)
	36 Mbps (16 QAM, R_c = 3/4)
	48 Mbps (64 QAM, R_c = 2/3)
	54 Mbps (64 QAM, R_c = 3/4)
Multiple access method	CSMA/CA

Table 8.6
HIPERLAN/2 and MMAC Parameters

Channel spacing	20 MHz
Bandwidth	16.56 MHz (–3 dB)
Number of subcarriers	52
Number of pilot subcarriers	4
Useful symbol length (t_s)	3.2 μs
Subcarrier separation (Δf)	312.5 kHz
Guard interval length (Δ_G)	800 ns
FEC	Convolutional code
Interleaving	Frequency domain bit interleaving (within one OFDM symbol)
Information transmission rate/ Modulation/coding rate	6 Mbps (BSPK, $R_c = 1/2$) 9 Mbps (BSPK, $R_c = 3/4$) 12 Mbps (QSPK, $R_c = 1/2$) 18 Mbps (QSPK, $R_c = 3/4$) 27 Mbps (16 QAM, $R_c = 9/16$) 36 Mbps (16 QAM, $R_c = 3/4$) 54 Mbps (64 QAM, $R_c = 3/4$)
Multiple access method	TDMA/DSA

8.4 Others

8.4.1 IEEE 802.11g

The IEEE 802.11b standard now supports 11-Mbps data transmission in the 2.4-GHz band [10], which is also called "industrial, scientific, and medical (ISM) band." Extensions up to 56-Mbps data transmission has been discussed in the ISM band as IEEE 802.11g. The IEEE 802.11g standard is still under development in July 2002, but will be finished by the end of 2002. The use of OFDM is already decided, and the same PHY as the IEEE 802.11a is likely to be used [11].

8.4.2 IEEE 802.11h

In Europe, 5.15- to 5.35-GHz and 5.45- to 5.725-GHz bands (for a total of 455-MHz bandwidth) can be allocated for HIPERLANs, but some bands require transmission power control (TPC) and dynamic frequency selection (DFS) to coexist with radar systems. Therefore, the IEEE 802.11a is not directly applicable. To make an IEEE 802.11a-based wireless LAN system available in Europe, the IEEE 802.11 standardization group is discussing a

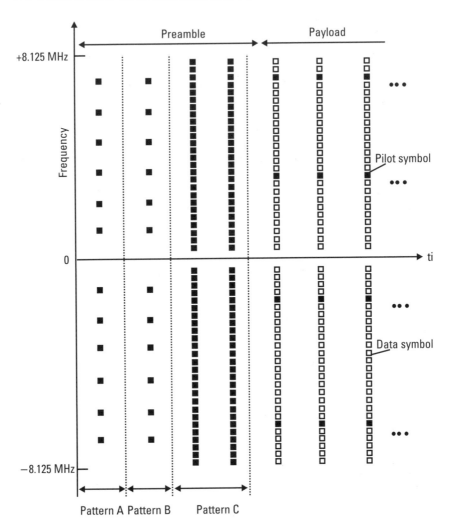

Figure 8.2 Time-frequency structure of an IEEE 802.11a-based OFDM packet.

new standard called IEEE 802.11h, which is an extension of the IEEE 802.11a with TPC and DFS [11].

8.4.3 IEEE 802.16a

IEEE 802.16 defines the WirelessMAN^{TM} air interface specification for metropolitan area networks (MANs), which attempts to replace "the last mile wired access" with cable modem and digital subscriber lines (DSL) by broadband wireless access (BWA). IEEE 802.16a standards, which will be finished by the end of 2002, define the air interface in 2- to 11-GHz band, including both licensed and license-exempt spectra. The IEEE 802.16a draft includes the following three specifications [12, 13]:

- WirelessMAN-SC2: This uses a single-carrier modulation format.
- WirelessMAN-OFDM: This uses an OFDM format with 256 subcarriers, and the access is based on TDMA.
- WirelessMAN-OFDMA: This uses an OFDM access (OFDMA) with 2,048 subcarriers.

This system performs multiple access by allocating a subset of the multiple subcarriers to an individual receiver. This system also uses frequency hopping (FH) spread spectrum (SS) for interference suppression.

8.5 Conclusions

As discussed in this chapter, through the research and development of digital broadcasting systems and wireless LAN systems, we have a lot of know-how on implementing OFDM. In Chapters 9 and 10, we will discuss the modification and application of OFDM for realization of 4G systems.

References

[1] Dambacher, P., "Digital Broadcasting," London: IEE, 1996.

[2] ETSI 300 744, "Digital Broadcasting Systems for Television, Sound, and Data Services; Framing Structure, Channel Coding and Modulation for Digital Terrestrial Television," ETSI, 1996.

[3] ARIB STD-B24, "Data Coding and Transmission Specification for Digital Broadcasting," ARIB, June 2000.

[4] IEEE Std. 802.11a, "Wireless Medium Access Control (MAC) and Physical Layer (PHY) Specifications: High-speed Physical Layer Extension in the 5-GHz Band," IEEE, 1999.

[5] ETSI TR 101 475, "Broadband Radio Access Networks (BRAN); HIPERLAN Type 2; Physical (PHY) Layer," ETSI BRAN, 2000.

[6] ARIB STD-T70, "Lower Power Data Communication Systems Broadband Mobile Access Communication System (CSMA)," ARIB, December 2000.

[7] ARIB STD-T70, "Lower Power Data Communication Systems Broadband Mobile Access Communication System (HiSWANa)," ARIB, December 2000.

[8]   van Nee, R., and R. Prasad, *OFDM for Wireless Multimedia Communications,* Norwood, MA: Artech House, 2000.

[9]   van Nee, R., et al., "New High-Rate Wireless LAN Standards," *IEEE Commun. Mag.,* Vol. 37, No. 12, December 1999, pp. 82–88.

[10]   IEEE Std. 802.11b, "Wireless Medium Access Control (MAC) and Physical Layer (PHY) Specifications: High-Speed Physical Layer Extension in the 2.4-GHz Band," IEEE, 1999.

[11]   Morikura, M., and H. Matsue, "Trends for IEEE 802.11-Based Wireless LAN," *IEICE Trans.,* Vol. J84-B, No. 11, November 2001, pp. 1918–1927.

[12]   Koffman, I., and V. Roman, "Broadband Wireless Access Solution Based on OFDM Access in IEEE 802.16," *IEEE Commun. Mag.,* Vol. 40, No. 4, April 2002, pp. 96–103.

[13]   Eklund, C., et al., "IEEE Standard 802.16: WirelessMANTM Air Interface for Broad-Band Wireless Access," *IEEE Commun. Mag.,* Vol. 40, No. 6, June 2002, pp. 98–107.

9

Combination of OFDM and CDMA

9.1 Introduction

It is well known that the CDMA scheme is robust to frequency selective fading and has been successfully introduced in commercial cellular mobile communications systems such as IS-95 and 3G systems [1–3]. On the other hand, as shown in Chapter 4, the OFDM scheme is also inherently robust to frequency selective fading. Therefore, no one would expect any synergistic effect in combination of the OFDM and CDMA schemes.

In 1993, the MC-CDM/CDMA system, which is indeed a combination of the two schemes, was independently proposed by three different groups [4–6]. So far, the MC-CDM/CDMA system has drawn a lot of attention, and we have conducted intensive research on this interesting system [7–18]. Now, in 2002, the MC-CDM/CDMA system is considered to be one of candidates as a physical layer protocol for 4G mobile communications, because 4G systems require high scalability and adaptability in the possible transmission rate and the MC-CDMA has the potential.

This chapter discusses the MC-CDMA system in detail. Section 9.2 modifies the channel model introduced in Chapter 2 to discuss the BER performance of a multiplexing system in downlink and a multiple access system in uplink. Section 9.3 presents the principle of a direct sequence (DS)-CDMA system and analyzes the BER lower bound. Section 9.4 discusses the MC-CDMA system. The section presents the principle of the MC-CDMA system with four singleuser combining schemes in downlink and two multiuser detection schemes in uplink, shows a head/tail guard interval insertion method not only to eliminate ISI and multiple user interference (MUI) but also to have virtually synchronous signal reception in uplink, compares the BER performance between the DS-CDMA and MC-CDMA systems, and shows a sliding DFT-based subcarrier recovery method. Finally, Section 9.5 offers some conclusions.

Strictly speaking, we should use "multiplexing" when referring to the downlink where several signals are multiplexed at a base station and use "multiple access" for uplink where several signals access a common wireless channel. However, we often see the word "CDMA in downlink" or "CDMA downlink." In this chapter, we will use the original wording and try not to be misleading.

9.2 Channel Model

As a frequency selective fading channel, we assume a WSSUS channel with L received paths in the complex equivalent baseband impulse response:

$$h_j(\tau;\ t) = \sum_{l=1}^{L} \beta_{l,j}(t)\,\delta(\tau - \tau_{l,j}) \qquad (9.1)$$

where j is the index for the user, $\beta_{l,j}(t)$ is the lth path gain, and $\tau_{l,j}$ is the propagation delay for the lth path. Assuming an identical and independent channel for an individual user, where the lth path is a mutually independent complex Gaussian variable with an average of zero and variance of σ_l^2, the multipath delay profile of the channel is given by

$$\phi_{h,j}(\tau) = \phi_h(\tau) = \sum_{l=1}^{L} \sigma_l^2\,\delta(\tau - \tau_l) \qquad (9.2)$$

and the spaced-frequency correlation function of the channel is given by

$$\phi_H(\Delta f) = \int_{-\infty}^{+\infty} \phi_h(\tau)\,e^{-j2\pi\Delta f \tau}\,d\tau \qquad (9.3)$$

Figure 9.1 shows the multipath delay profile.

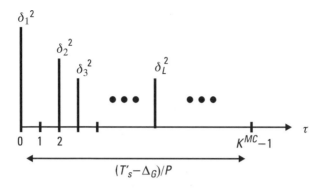

Figure 9.1 Multipath delay profile of the channel for the *j*th user.

9.3 DS-CDMA System

9.3.1 DS-CDMA Transmitter

Figure 9.2(a) shows the DS-CDMA transmitter for the *j*th user with coherent PSK format. The complex equivalent baseband transmitted signal is written as

$$s_j^{DS}(t) = \sum_{i=-\infty}^{+\infty} \sum_{k=0}^{K^{DS}-1} a_j(i) \, b_j(k) \, c_j(k + iK^{DS}) \, p_c(t - kT_c - iT_s)$$

$$(9.4)$$

where $a_j(i)$, $b_j(k)$, and $c_j(k)$ are the *i*th information symbol, the *k*th chip of the short spreading (channelizing) code with length K^{DS} and *k*th chip of the long spreading (scrambling) code with much longer length than K^{DS}, respectively. Here, the two spreading codes are normalized as follows:

$$\sum_{k=0}^{K^{DS}-1} |b_j(k)|^2 = 1 \qquad (9.5)$$

$$\sum_{k=0}^{K^{DS}-1} |c_j(k)|^2 = 1 \qquad (9.6)$$

Furthermore, in (9.4), T_c and T_s are the chip duration and symbol duration, respectively, and $p_c(t)$ is the chip pulse waveform.

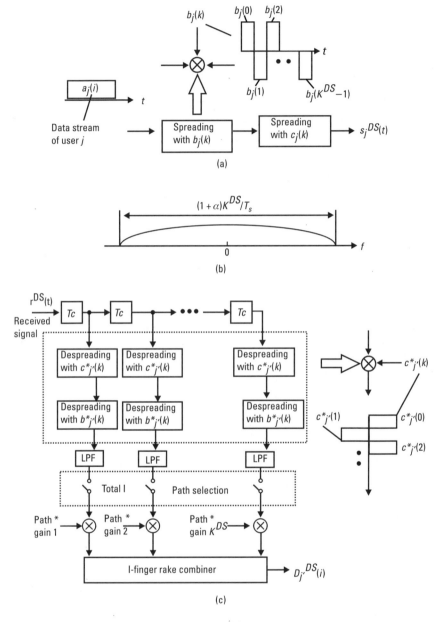

Figure 9.2 DS-CDMA system: (a) transmitter; (b) power spectrum of transmitted signal; and (c) *I*-finger Rake receiver.

Figure 9.2(b) shows the power spectrum of the transmitted signal. If employing the Nyquist filter with roll-off factor of α as a baseband pulse shaping filter, the bandwidth is given by

$$B^{DS} = (1 + \alpha) K^{DS}/T_s \qquad (9.7)$$

9.3.2 DS-CDMA Receiver

Figure 9.2(c) shows the DS-CDMA receiver with the I-finger Rake combiner for the j'th user. The received wave through the channel given by (9.1) is written as

$$r^{DS}(t) = \sum_{j=1}^{J} r_j^{DS}(t) + n(t) \qquad (9.8)$$

$$r_j^{DS}(t) = \int_{-\infty}^{+\infty} s_j^{DS}(t - \tau) h_j(\tau; t) \, d\tau = \sum_{l=1}^{J} \beta_{l,j}(t) s_j^{DS}(t - \tau_l) \qquad (9.9)$$

where J is the number of total active users, $r_j^{DS}(t)$ is the received signal component for the jth user, and $n(t)$ is the complex additive Gaussian noise component with an average of zero and variance of σ_n^2.

The decision variable for the j'th user at $t = iT_s$ is written as

$$D_{j'}^{DS}(i) = D_{j'}^{DS}(t = iT_s)$$

$$= \sum_{l=1}^{I} \beta_{l,j'}^{*}(iT_s) \frac{1}{T_s} \int_{iT_s + \tau_l}^{(i+1)T_s + \tau_l} b_{j'}^{*}(k) c_{j'}^{*}(k + iK^{DS}) \qquad (9.10)$$

$$\times p_c(t - kT_c - iT_s - \tau_l) r^{DS}(t) \, dt$$

The symbol decision on $a_j(i)$ is made based on the polarity of the inphase and quadrature components of (9.10). We define the decision process as

$$\hat{a}_j(i) = \text{DEC}\left[r^{DS}(t) \right] \qquad (9.11)$$

The Rake combiner is a kind of single user detection scheme that requires no information on other active users. Therefore, not only can a mobile terminal in a downlink (from a base station to an individual user) but also a base station in an uplink (from an individual user to a base station) can use the Rake combiner. On the other hand, in uplink, a base station can know information on all active users. Therefore, it can also use a multiuser detection scheme.

At a base station, the (estimated) channel impulse response and detected information symbol for all active users are available, so the base station can generate the replica of the received signal for the jth user:

$$\hat{r}_j^{DS}(t) = \int_{-\infty}^{+\infty} \hat{s}_j^{DS}(t-\tau)h_j(\tau;t)\,d\tau = \sum_{l=1}^{I} \beta_{l,j}(t)\hat{s}_j^{DS}(t-\tau_l)$$

(9.12)

$$\hat{s}_j^{DS}(t) = \sum_{i=-\infty}^{+\infty} \sum_{k=0}^{K^{DS}-1} \hat{a}_j(i)b_j(k)c_j(k+iK^{DS})p_c(t-kT_c-iT_s)$$

(9.13)

The serial interference cancellation (SIC) scheme first reorders the active users in a decreasing order with respect to the received signal power. Defining the received signal power for the qth user as

$$P_q = \frac{1}{2}E\left[|r_q^{DS}(t)|^2\right]$$

(9.14)

the received wave can be rewritten as

$$r^{DS}(t) = \sum_{q=1}^{J} r_q^{DS}(t) + n(t)$$

(9.15)

with

$$P_{q-1} \geq P_q, \quad (q = 2, 3, \ldots, J)$$

(9.16)

Taking into consideration that the received signal with a larger power could have a higher reliability for demodulation, the SIC scheme then makes the symbol decision as

$$\hat{a}_1(i) = \text{DEC}\left[r^{DS}(t)\right] \tag{9.17}$$

$$\hat{a}_{q'}(i) = \text{DEC}\left[r^{DS}(t) - \sum_{q=1}^{q'-1} \hat{r}_j^{DS}(t)\right], \ (q' = 2, 3, \ldots, J)$$

9.3.3 Bit Error Rate Analysis

Assume a single user case. Defining $\mathbf{r}_t$ as the $(L \times 1)$ received signal vector:

$$\mathbf{r}_t = [r_1, r_2, \ldots r_L]^T \tag{9.18}$$

the $(L \times L)$ time domain covariance matrix $\mathbf{R}_t$ for the received signal vector is given by

$$\mathbf{R}_t = \frac{1}{2} E\left[\mathbf{r}_t \cdot \mathbf{r}_t^H\right] = \begin{bmatrix} \sigma_1^2 & 0 & \cdots & 0 \\ 0 & \sigma_2^2 & \ddots & \vdots \\ \vdots & \ddots & \ddots & 0 \\ 0 & \cdots & 0 & \sigma_L^2 \end{bmatrix} \tag{9.19}$$

In (9.19), we assume a perfect autocorrelation characteristic for the spreading codes.

The BER of time domain I-finger DS-CDMA Rake combiner for the case of a single user is uniquely determined by the eigenvalues of $\mathbf{R}_t$ (in this case, the eigenvalues are clearly $\sigma_1^2, \ldots, \sigma_L^2$) [19].

For example, when σ_l^2 $(l = 1, \ldots, L)$ are different from each other, the BER is given by

$$BER^{DS} = \sum_{l=1}^{I} w_l \frac{1}{2} \left(1 - \sqrt{\frac{\sigma_I^2/\sigma_n^2}{1 + \sigma_I^2/\sigma_n^2}}\right) \tag{9.20}$$

$$w_l = \frac{1}{\displaystyle\prod_{\substack{v=1 \\ v \neq l}}^{I} (1 - \sigma_v^2/\sigma_l^2)} \tag{9.21}$$

$$\sigma_{total}^2 = \sum_{l=1}^{L} \sigma_l^2 \qquad (9.22)$$

where σ_{total}^2 is the total power of the received signal.

Also, when σ_l^2 ($l = 1, \ldots, L$) are all the same ($= \sigma_s^2$)

$$BER^{DS} = \left(\frac{1 - \mu^{DS}}{2}\right)^I \sum_{l=0}^{I-1} \binom{I-1+l}{l} \left(\frac{1 - \mu^{DS}}{2}\right)^l \qquad (9.23)$$

$$\mu^{DS} = \sqrt{\frac{\sigma_s^2/\sigma_n^2}{1 + \sigma_s^2/\sigma_n^2}} \qquad (9.24)$$

Note that the L-finger Rake combiner can achieve the minimum BER (the BER lower bound).

9.4 MC-CDMA System

9.4.1 MC-CDMA Transmitter

The OFDM scheme is insensitive to frequency selective fading but it has severe disadvantages such as difficulty in subcarrier synchronization and sensitivity to frequency offset and nonlinear amplification; on the other hand, the CDMA scheme has robustness against frequency selective fading. Therefore, any synergistic effect might not be expected in combining an OFDM scheme with a CDMA scheme. However, the combination has two major advantages. One is its own capability to lower the symbol rate in each subcarrier enough to have a quasi-synchronous signal reception in uplink. The other is that it can effectively combine the energy of the received signal scattered in the frequency domain. Especially for high-speed transmission cases where a DS-CDMA receiver could see 20 paths in the instantaneous impulse response, a 20-finger Rake combiner would be impossible to implement for the DS-CDMA receiver, whereas an MC-CDMA receiver would be possible although it would lose the energy of the received signal in the guard interval.

An MC-CDMA transmitter spreads the original signal using a given spreading code in the frequency domain. In other words, a fraction of the symbol corresponding to a chip of the spreading code is transmitted through a different subcarrier. For multicarrier transmissions, it is essential to have frequency nonselective fading over each subcarrier. Therefore, if the original symbol rate is high enough to become subject to frequency selective fading, the signal needs to first be converted from serial to parallel before spreading over the frequency domain.

The basic transmitter structure of an MC-CDMA scheme is similar to that of an OFDM scheme. The main difference is that the MC-CDMA scheme transmits the same symbol in parallel through different subcarriers, whereas the OFDM scheme transmits different symbols.

Figure 9.3(a) shows the MC-CDMA transmitter for the jth user with CPSK format. The input information sequence is first converted into P parallel data sequences $(a_{j,0}(i), a_{j,1}(i), \ldots, a_{j,P-1}(i))$ and each serial/parallel-converter output is multiplied with the short spreading code $d_j(m)$ with length K^{MC}. The P parallel data sequences are converted back to a serial data sequence, and the resultant data sequence is again multiplied with the long spreading code $c_j(m)$ with much longer length than K^{MC}. The spread data is then mapped onto PK^{MC} subcarriers through the PK^{MC}-point IDFT, and finally the guard interval Δ_G is inserted between OFDM symbols to avoid ISI caused by multipath fading. The complex equivalent baseband transmitted signal is written as

$$s_j^{MC}(t) = \sum_{i=-\infty}^{+\infty} \sum_{p=0}^{P-1} \sum_{m=0}^{K^{MC}-1} a_{j,p}(i) d_j(m) c_j(Pm + p + iPK^{MC}) \quad (9.25)$$

$$\times p_s(t - iT_s') e^{j2\pi(Pm+p)\Delta f'(t-iT_s')}$$

$$T_s' = \Delta_G + t_s = PT_s \quad (9.26)$$

$$\Delta f' = 1/(T_s' - \Delta_G) \quad (9.27)$$

where $d_j(m)$ and $c_j(m)$ are normalized as

$$\sum_{m=0}^{K^{DS}-1} |d_j(m)|^2 = 1 \quad (9.28)$$

$$\sum_{m=0}^{K^{DS}-1} |c_j(m)|^2 = 1 \quad (9.29)$$

Furthermore, in (9.25) through (9.27), T_s' is the symbol duration at subcarrier level, $\Delta f'$ is the subcarrier separation, and $p_s(t)$ is the rectangular symbol pulse waveform defined as

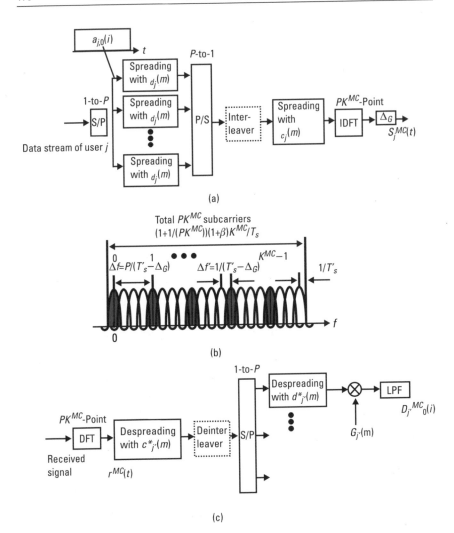

Figure 9.3 MC-CDMA system: (a) transmitter; (b) power spectrum of transmitted signal; and (c) receiver.

$$p_s(t) = \begin{cases} 1, & (-\Delta_G \leq t < t_s) \\ 0, & (\text{otherwise}) \end{cases} \qquad (9.30)$$

Figure 9.3(b) shows the power spectrum of the transmitted signal. The bandwidth of the transmitted signal is written as

$$B^{MC} = (PK^{MC} - 1)/(T_s' - \Delta_G) + 2/T_s'$$

$$\approx (PK^{MC} + 1)/(T_s' - \Delta_G) \tag{9.31}$$

$$= (1 + 1/(PK^{MC}))/(1 - \Delta_G/(PT_s)) K^{MC}/T_s$$

$$\approx (1 + 1/(PK^{MC}))(1 + \gamma) K^{MC}/T_s$$

$$\gamma = \Delta_G/(PT_s) \tag{9.32}$$

where γ is the bandwidth expansion factor associated with the guard interval insertion.

Equation (9.25) shows that no spreading operation is done in the time domain and that the symbol duration at subcarrier level is P times as long as the original symbol level because of the serial/parallel conversion. Furthermore, let us neglect the interleaver in Figure 9.3(a) and the corresponding deinterleaver in Figure 9.3(c). (The effect of the interleaver and deinterleaver will be discussed in Section 10.3.) Then, the subcarrier separation for $a_{j,p}(i)$ becomes $P\Delta f'$ [see the hatched subcarrier power spectra in Figure 9.3(b)], therefore, full frequency diversity effect is expected in the MC-CDMA system.

9.4.2 MC-CDMA Receiver

Figure 9.3(c) shows the MC-CDMA receiver for the j'th user. The received wave is written as [9, 11, 14]

$$r^{MC}(t) = \sum_{j=1}^{J} r_j^{MC}(t) + n(t) \tag{9.33}$$

$$r_j^{MC}(t) = \int_{-\infty}^{+\infty} s_j^{MC}(t - \tau) h_j(\tau; t) \, d\tau$$

$$= \sum_{i=-\infty}^{+\infty} \sum_{p=0}^{P-1} \sum_{m=0}^{K^{MC}-1} z_{m,p,j}(t) a_{j,p}(i) d_j(m) c_j(Pm + p + iPK^{MC})$$

$$\times p_s(t - iT_s') e^{j2\pi(Pm+p)\Delta f'(t-iT_s')} \tag{9.34}$$

where $z_{m,p,j}(t)$ is the received complex envelope at the $(Pm + p)$th subcarrier of the jth user.

The received wave is first fed into the PK^{MC}-point DFT and the Fourier coefficients, namely, the complex envelope values, are calculated for all the PK^{MC} subcarriers. After despreading with the long and short codes, the mth subcarrier components are multiplied with $G_{j'}(m)$ to combine the energy of the received signal scattered in the frequency domain.

The decision variable for the jth user at $t = iT_s$ is the sum of the weighted baseband components written as

$$D_{j',p'}^{MC}(i) = D_{j',p'}^{MC}(t = iT_s')$$

$$= \sum_{m=0}^{K^{MC}-1} d_{j'}^*(m) c_{j'}^*(Pm + p' + iPK^{MC}) G_{j',p',i}(m) y_{p',i}(m)$$

(9.35)

$$y_{p',i}(m) = \sum_{j=1}^{J} z_{m,p',j}(iT_s') a_{j,p'}(i) d_j(m) c_j(Pm + p' + iPK^{MC})$$

$$+ n_{m,p'}(iT_s')$$

(9.36)

where $y_{p',i}(m)$ and $n_{m,p'}(iT_s')$ are the complex baseband component of the received signal and the complex additive Gaussian noise at the $Pm + p'$th subcarrier at $t = iT_s'$, respectively.

Now, we discuss the following four combining schemes in the downlink and two multiuser detection schemes in the uplink.

In the downlink ($z_{m,p',1} = z_{m,p',2} = \cdots z_{m,p',J} = z_{m,p'}$) where we can drop the subscript j, orthogonality restoring combining (ORC) chooses the gain as

$$G_{j',p',i}(m) = z_{m,p'}^*(iT_s')/|z_{m,p'}(iT_s')|^2$$

(9.37)

so the receiver can eliminate multiple access interference perfectly:

$$\hat{a}_{j',p'}(i) = D_{j',p'}^{MC}(i) = a_{j',p'}(i) + \sum_{m=0}^{K^{MC}-1} z_{m,p'}^*(iT_s')/|z_{m,p'}(iT_s')|^2$$

$$\times d_j^*(m) c_j^*(Pm + p' + iK^{MC}) n_{m,p'}(iT_s')$$

(9.38)

In (9.38), low-level subcarriers tend to be multiplied with high gains and the noise components are amplified at weaker subcarriers. This noise amplification effect degrades the BER performance.

Equal gain combining (EGC) chooses the gain as

$$G_{j',p',i}(m) = z^*_{m,p'}(iT'_s)/|z_{m,p'}(iT'_s)| \qquad (9.39)$$

and maximum ratio combining (MRC) chooses the gain as

$$G_{j',p',i}(m) = z^*_{m,p'}(iT'_s) \qquad (9.40)$$

In the case of a single user, the MRC can minimize the BER.
Finally, a problem for minimization of the mean square error is given by

$$\text{minimize MSE}(G_{j',p',i}(m)) = E|a_{j',p'}(i) - \hat{a}_{j',p'}(i)|^2 \qquad (9.41)$$

According to the principle of orthogonality, the error must be orthogonal to all the baseband components of the received subcarriers:

$$E[(a_{j',p'}(i) - \hat{a}_{j',p'}(i))y_{p',i}(m)] = 0, \qquad (m = 0, 1, \ldots K^{MC} - 1) \qquad (9.42)$$

The solution of (9.42) gives the gain of minimum mean square error combining (MMSEC) as

$$G_{j',p',i}(m) = z^*_{m,p'}(iT'_s)/(J|z_{m,p'}(iT'_s)|^2 + \sigma^2_n) \qquad (9.43)$$

Note that, in the downlink, for small $|z_{m,p'}|$, the gain becomes small to avoid excessive noise amplification, whereas for large $|z_{m,p'}|$, it becomes in proportion to the inverse of the subcarrier envelope $z^*_{m,p'}/|z_{m,p'}|^2$ to recover orthogonality among users.

The decision variable on the transmitted symbol is given by (9.9) for a DS-CDMA system, whereas it is given by (9.35) for the MC-CDMA system, and the two equations clearly show one of advantages of the MC-CDMA system, that is, for the symbol decision, the DS-CDMA system requires a kind of complicated convolution, whereas the MC-CDMA system requires just a multiplication, namely, a one-tap equalizer for each subcarrier. This is also clear from the fact that the Fourier transform for convolution of two-time domain-functions is given by the multiplication of two frequency domain functions obtained through the Fourier transform of the two-time domain functions.

Next, to discuss multiuser detection schemes in the uplink, for the sake of simplicity and without loss of generality, let us drop the subscripts i and p' in (9.36). Linear multiuser detection scheme, which means a detection using a linear sum of the received waves, is defined as [15, 16]

$$\hat{a}_{j'} = \mathrm{DEC}\left[\mathbf{w}_{j'}^{H}\mathbf{y}\right] \tag{9.44}$$

where $\mathbf{w}_{j'}$ is the $(K^{MC} \times 1)$ weight vector for the j'th user to be determined:

$$\mathbf{w}_{j'} = [w_{j',0}, \ldots w_{j',K^{MC}-1}]^{T} \tag{9.45}$$

and $\mathbf{y}$ is the $(K^{MC} \times 1)$ received wave vector. For a quasi-synchronous MC-CDMA uplink, $\mathbf{y}$ is the $(K^{MC} \times 1)$ Fourier coefficient vector defined as

$$\mathbf{y} = [y_{p'}(0), \ldots y_{p'}(K^{MC} - 1)]^{T} \tag{9.46}$$

To write (9.36) in a vector form, define the $(K^{MC} \times 1)$ distorted spreading code vector for the jth user, the $(J \times 1)$ transmitted symbol vector and the $(K^{MC} \times 1)$ noise vector as

$$\mathbf{d}_{j} = [d_{j,0}, \ldots d_{j,K^{MC}-1}]^{T} \tag{9.47}$$

$$d_{j,m} = z_{m,p',j}(iT_{s}')d_{j}(m)c_{j}(Pm + p' + iK^{MC})$$

$$\mathbf{a} = [a_{1,p'}(i), \ldots, a_{J,p'}(i)]^{T} \tag{9.48}$$

$$\mathbf{n} = [n_{0,p'}(iT_{s}'), \ldots, n_{K^{MC}-1,p'}(iT_{s}')]^{T} \tag{9.49}$$

furthermore, define the $(K^{MC} \times J)$ distorted spreading code matrix as

$$\mathbf{D} = [\mathbf{d}_{1}, \ldots, \mathbf{d}_{J}] \tag{9.50}$$

Using (9.46) to (9.50), (9.36) can be written as

$$\mathbf{y} = \mathbf{D}\mathbf{a} + \mathbf{n} \tag{9.51}$$

The received signal may be despread with the distorted spreading codes:

$$\mathbf{D}^H \mathbf{y} = \mathbf{D}^H \mathbf{D} \mathbf{a} + \mathbf{D}^H \mathbf{n} \tag{9.52}$$

However, DHD in (9.52) cannot be the ($J \times J$) identity matrix, because the orthogonality among the spreading codes is totally distorted through the frequency selective fading channel.

The decorrelating multiuser detection scheme eliminates the crosscorrelation among the spreading codes by multiplying (9.52) with $(\mathbf{D}^H \mathbf{D})^{-1}$:

$$(\mathbf{D}^H \mathbf{D})^{-1} \mathbf{D}^H \mathbf{y} = \mathbf{a} + (\mathbf{D}^H \mathbf{D})^{-1} \mathbf{D}^H \mathbf{n} \tag{9.53}$$

Therefore, the weight vector of the decorrelating multiuser detection scheme for the j'th user is given by

$$\mathbf{w}_{j'}^{dec} = \sum_{j=1}^{J} [(\mathbf{D}^H \mathbf{D})^{-1}]_{j',j}\, \mathbf{d}_j \tag{9.54}$$

where $[\mathbf{A}]_{j',j}$ means the (j', j) element of matrix $\mathbf{A}$.

On the other hand, the MMSE multiuser detection scheme minimizes the following mean square error:

$$\text{minimize MSE}(\mathbf{w}_{j'}) = E\left[\left(a_{j'} - \mathbf{w}_{j'}^H \mathbf{y}\right)^2\right] \tag{9.55}$$

The mean square error can be written as

$$\begin{aligned}\text{MSE}(\mathbf{w}_{j'}) &= 1 - 2\mathbf{w}_{j'}^H E[a_{j'}\mathbf{y}] + \mathbf{w}_{j'}^H E[\mathbf{y}\mathbf{y}^H]\mathbf{w}_{j'} \\ &= 1 - 2\mathbf{w}_{j'}^H \mathbf{d}_{j'} + 2\mathbf{w}_{j'}^H \mathbf{Y}\mathbf{w}_{j'}\end{aligned} \tag{9.56}$$

where $E[a_{j'}\mathbf{y}] = \mathbf{d}_{j'}$ and $\mathbf{Y} = \frac{1}{2}E[\mathbf{y}\mathbf{y}^H]$ is the correlation matrix of the received wave vector.

From $\partial\, \text{MSE}(\mathbf{w}_{j'})/\partial \mathbf{w}_{j'} = 0$, the weight vector of the MMSE multiuser detection scheme for the j'th user is given by

$$\mathbf{Y}\mathbf{w}_{j'} = \mathbf{d}_{j'} \tag{9.57}$$

$$\therefore\ \mathbf{w}_{j'} = \mathbf{Y}^{-1}\mathbf{d}_{j'}$$

9.4.3 Bit Error Rate Analysis

Assume a single user case [14]. Defining $\mathbf{r}_f$ as the ($K^{MC} \times 1$) received signal vector, the ($K^{MC} \times K^{MC}$) frequency domain covariance matrix $\mathbf{R}_f$ is given by

$$\mathbf{r}_f = [z_{0,p'}, \ldots, z_{K^{MC}-1,p'}]^T \tag{9.58}$$

$$\mathbf{R}_f = \frac{1}{2}[\mathbf{r}_f \mathbf{r}_f^H] = \{m_{a,b}\} \tag{9.59}$$

$$m_{a,b} = \phi_H((a-b)P\Delta f')$$

where $\{m_{a,b}\}$ is the matrix with (a, b) element $m_{a,b}$ and $\phi_H(\Delta f)$ is given by (9.3).

Defining $\lambda_0, \ldots, \lambda_{K^{MC}-1}$ as the nonzero eigenvalues of $\mathbf{R}_f$, the BER is given by a form similar to (9.20) or (9.23). For example, when λ_n $(n = 0, \ldots, K^{MC} - 1)$ are different from each other, the BER is given by

$$BER^{MC} = \sum_{n=0}^{K^{MC}-1} v_n \frac{1}{2}\left(1 - \sqrt{\frac{\lambda_n/\sigma_n^2}{1 + \lambda_n/\sigma_n^2}}\right) \tag{9.60}$$

$$v_n = \frac{1}{\displaystyle\prod_{\substack{u=0 \\ u \neq n}}^{K^{MC}-1} (1 - \lambda_u/\lambda_n)} \tag{9.61}$$

Also, when λ_n $(n = 0, \ldots, K^{MC} - 1)$ are all the same $(= \lambda)$

$$BER^{MC} = \left(\frac{1 - \mu^{MC}}{2}\right)^{K^{MC}} \sum_{n=0}^{K^{MC}-1} \binom{K^{MC} - 1 + n}{n} \left(\frac{1 + \mu^{MC}}{2}\right)^n \tag{9.62}$$

$$\mu^{MC} = \sqrt{\frac{\lambda/\sigma_n^2}{1 + \lambda/\sigma_n^2}} \tag{9.63}$$

We can show that (9.20) is equivalent to (9.60) and that (9.23) is equivalent to (9.62) as follows.

For the multipath delay profile shown in Figure 9.1, we can define the following $(K^{MC} \times K^{MC})$ time domain covariance matrix with time resolution of $(T_s' - \Delta_G)/(PK^{MC})$:

$$\mathbf{R}'_t = \begin{bmatrix} \sigma_1^2 & 0 & \cdots & \cdots & 0 \\ 0 & 0 & \ddots & & \vdots \\ \vdots & \ddots & \sigma_2^2 & \ddots & \vdots \\ \vdots & & \ddots & \sigma_3^2 & 0 \\ 0 & \cdots & \cdots & 0 & \ddots \end{bmatrix} \tag{9.64}$$

where the nonzero eigenvalues of $\mathbf{R}'_t$ are $\sigma_1^2, \ldots, \sigma_L^2$.

The corresponding $(K^{MC} \times K^{MC})$ frequency domain covariance matrix with frequency resolution of $P/(T'_s - \Delta_G)$ is given by

$$\mathbf{R}'_f = \mathbf{W}(K^{MC}) \mathbf{R}'_t \mathbf{W}^H(K^{MC}) \tag{9.65}$$

where $\mathbf{W}(K^{MC})$ is the normalized $(K^{MC} \times K^{MC})$ DFT matrix given by

$$\mathbf{W}(K^{MC}) = \{w_{a,b}\} \tag{9.66}$$

$$w_{a,b} = \frac{1}{\sqrt{K^{MC}}} e^{j2\pi(ab/K^{MC})}$$

with the following property:

$$\mathbf{W}^{-1}(K^{MC}) = \mathbf{W}^H(K^{MC}) \tag{9.67}$$

Define $\mathbf{r}_l$ as the $(K^{MC} \times 1)$ eigenvector associated with the eigenvalue σ_l^2:

$$\mathbf{R}'_t \mathbf{r}_l = \sigma_l^2 \quad \mathbf{r}_l, \ (l = 1, 2, \ldots, L) \tag{9.68}$$

and also define the $(K^{MC} \times 1)$ vector $\mathbf{z}_l$ as

$$\mathbf{z}_l = \mathbf{W}(K^{MC}) \quad \mathbf{r}_l, \ (l = 1, 2, \ldots, L) \tag{9.69}$$

Now, we can theoretically prove that the eigenvalues of the frequency domain covariance matrix (9.65) are all the same as those of the time domain covariance (9.64):

$$\mathbf{R}'_f \mathbf{z}_l = \mathbf{W}(K^{MC})\,\mathbf{R}'_t\mathbf{W}^H(K^{MC}) \cdot \mathbf{W}(K^{MC})\,\mathbf{r}_l$$

$$= \mathbf{W}(K^{MC})\,\mathbf{R}'_t\,\mathbf{r}_l$$

$$= \mathbf{W}(K^{MC})\,\sigma_l^2\,\mathbf{r}_l \qquad\qquad (9.70)$$

$$= \sigma_l^2\,\mathbf{W}(K^{MC})\,\mathbf{r}_l$$

$$= \sigma_l^2\,\mathbf{z}_l$$

Equation (9.70) clearly shows that the nonzero eigenvalues of $\mathbf{R}'_f$ are $\sigma_1^2, \ldots, \sigma_L^2$. Therefore, as long as we assume the same frequency selective fading channel, the BER lower bound of the MC-CDMA system is all the same as that of the DS-CDMA system. Furthermore, the assumption of an independent fading characteristic at each subcarrier implies not a frequency nonselective fading but a frequency selective fading at each subcarrier, because it requires independent PK^{MC} paths uniformly scattered in the subcarrier level symbol duration, namely, $t_s = T'_s\text{-}\Delta_G$.

9.4.4 Design of MC-CDMA System

When the symbol transmission rate, channel frequency selectivity, and channel time selectivity are given, it is necessary in the MC-CDMA system to determine the number of subcarriers and the length of guard interval.

For a normal DPSK-based OFDM system, Section 4.5.3 theoretically discusses how to determine the optimum number of subcarriers and the optimum length of guard interval. It is also possible to theoretically discuss the same design issue for the MC-CDMA system, but in this section, we show some numerical results obtained from a straightforward computer simulation [17].

Table 9.1 shows the simulation parameters for the system design. We evaluate the BER performance in a single cell environment, so we neglect

Table 9.1
Simulation Parameters for System Design

Information transmission rate	4 [Mbps] (QPSK)
Short spreading codes	Walsh Hadamard codes with $K^{MC} = 32$
Combining scheme	MMSEC
Number of users	8
Channel fading	Frequency selective fast Rayleigh
Multipath delay profile	10-path exponentially decaying

the long spreading (scrambling) codes. Furthermore, we assume perfect channel parameter estimation and no channel coding.

Figure 9.4 shows the BER as a function of the number of subcarriers N_{SC} and Doppler frequency f_D, where we assume average $E_b/N_0 = 10$ dB, $\tau_{RMS} = 200$ nsec, and $\Delta_G = 10\%$. For the given delay spread, the system requires more than around 512 subcarriers. In addition, the system is sensitive to the Doppler shift, namely, the channel time variation. For the given parameter setting, it can keep a good BER up to $f_D = 200$ Hz.

Figure 9.5 shows the BER as a function of N_{SC} and τ_{RMS}, where we assume average $E_b/N_0 = 12$ dB, $f_D = 80$ Hz, and $\Delta_G = 5\%$. The system is sensitive to the delay spread. For the given parameter setting, it can keep a good BER up to $\tau_{RMS} = 10$ nsec.

Figure 9.6(a) shows the BER as a function of Δ_G and N_{SC} and Figure 9.6(b) shows N_{SC} against Δ_G for some given BERs, where we assume average $E_b/N_0 = 10$ dB, $f_D = 80$ Hz, and $\tau_{RMS} = 200$ nsec. From Figure 9.6(b), we can see that we could have two choices, namely, $N_{SC} = 1,024$ with $\Delta_G = 5\%$ and $N_{SC} = 512$ with $\Delta_G = 10\%$.

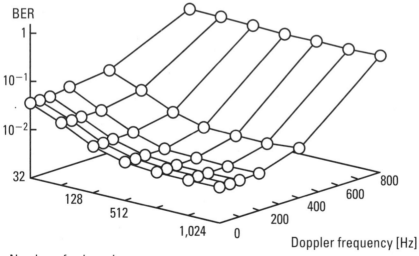

Figure 9.4 BER as a function of N_{SC} and f_D.

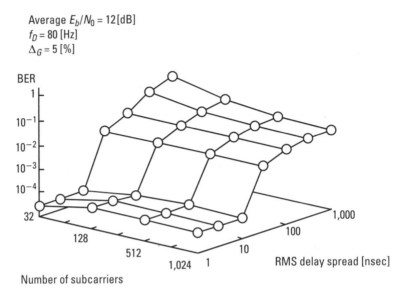

Average $E_b/N_0 = 12$ [dB]
$f_D = 80$ [Hz]
$\Delta_G = 5$ [%]

Figure 9.5 BER as a function of N_{SC} and τ_{RMS}.

Through several computer simulations, we found that the preferable number of subcarriers and preferable length of guard interval largely depends on the multipath delay profile chosen, such as the shape and the number of paths. Namely, a different multipath delay profile gives different values for optimum N_{SC} and Δ_G, so it is necessary to carefully choose the multipath delay profile with which we should evaluate the BER performance.

9.4.5 Head/Tail Guard Interval Insertion Method

Wireless channels cannot always have a minimum phase response, namely, a delayed path often has a less loss than the first path. In an example shown in Figure 9.7, the second path has the largest amplitude in the impulse response.

Correlation-type OFDM symbol timing synchronizers try to catch a path with the largest amplitude for DFT windowing. Therefore, when the channel has the nonminimum response, a problem arises for the conventional OFDM symbol format (head guard interval insertion). In Figure 9.7(a), ISI from the first path is included in the DFT window because the synchronizer catches the second path. For an MC-CDMA downlink, the transmitted signals for all the users are synchronized, so they are orthogonal among the same path but not orthogonal among different paths. Therefore, when this type of DFT window timing synchronizer is used for an MC-CDMA

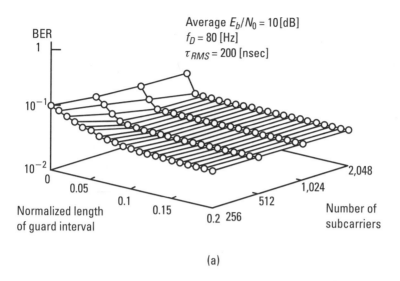

(a)

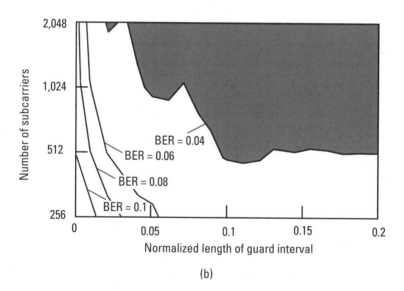

(b)

Figure 9.6 Preferable Δ_G and N_{SC}: (a) BER as a function of Δ_G and N_{SC}; and (b) Δ_G against N_{SC}.

downlink, not only ISI but also MUI deteriorate the BER when the channel has the nonminimum response.

Figure 9.7(b) shows a head/tail guard interval insertion method where one OFDM symbol is cyclically extended at both head and tail parts with

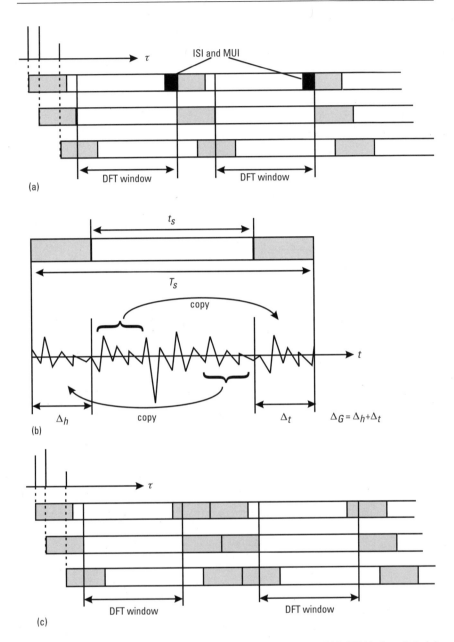

Figure 9.7 Effect of nonminimum phase response of channel in MC-CDMA downlink: (a) ISI and MUI in conventional head guard interval insertion; (b) head/tail guard interval insertion; and (c) no ISI and MUI in head/tail guard interval insertion.

lengths of Δ_h and Δ_t [17, 18]. Figure 9.7(c) shows the effect of the insertion method. This symbol format can cope with any channel response, with no ISI and no MUI.

The head/tail guard interval insertion method is much more effective in an MC-CDMA uplink. Figure 9.8 shows a quasi-synchronous MC-CDMA uplink. Even when signals from users can be quasi-synchronously received at a base station, for the conventional symbol format, ISI and MUI are included in the DFT window, which result in BER degradation [see Figure 9.8(a)]. On the other hand, when the head/tail guard interval insertion method is used, it can perfectly eliminate both ISI and MUI. Even when a quasi-synchronous signal transmission is employed in an MC-CDMA uplink, this insertion method can accomplish a virtually synchronous signal reception at the DFT windowing level.

9.4.6 Bit Error Rate of MC-CDMA System

In this section, we show numerical results on the BER performance of the MC-CDMA system obtained from computer simulation. For comparison purposes, we also show the BER performance of the DS-CDMA system. Here, we evaluate the BER in a single cell environment, so we neglect the long spreading codes. We assume perfect channel estimation and no channel coding, and, furthermore, we assume synchronous signal reception for MC-CDMA uplink by means of the head/tail guard interval insertion method. Table 9.2 shows the parameters common to the two systems, while Tables 9.3 and 9.4 show the parameters of MC-CDMA and DS-CDMA systems, respectively.

Figures 9.9 and 9.10 show the BER performance in the downlink for the MC-CDMA and DS-CDMA systems, respectively. In Figure 9.9, the MMSEC always outperforms the other three combining schemes such as ORC, MRC, and EGC, although it requires estimation of noise power. The performance of the ORC is very poor because of the noise enhancement. Taking the receiver complexity into consideration, the EGC may be a good choice, because it performs well and does not require noise power estimation. The best BER is achieved when there is only one user for the MRC, but there is still a large difference between the attainable lowest BER by the scheme and the lower bound. This is due to energy loss associated with the guard interval insertion (in this case, 10%-guard interval is assumed). If the guard interval length is shortened, the attainable lowest BER approaches to the lower bound, however, at the sacrifice of robustness against frequency selective fading. On the other hand, for the DS-CDMA system, in

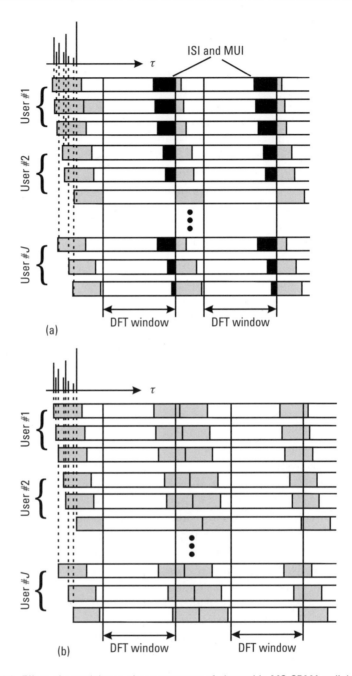

Figure 9.8 Effect of nonminimum phase response of channel in MC-CDMA uplink: (a) ISI and MUI in conventional head guard interval intrusion; and (b) no ISI and MUI in head/tail guard interval insertion.

Table 9.2
Simulation Parameters Common to MC-CDMA and DS-CDMA

Information transmission rate	512 [Kbps] (BPSK)
Channel fading	Frequency selective slow Rayleigh
Multipath delay profile	18-path exponentially decaying
Delay spread (τ_{RMS})	200 [nsec]
Average E_b/N_0 10 [dB]	

Table 9.3
Simulation Parameters for MC-CDMA

Number of subcarriers	9.24
Guard interval length	$\Delta_G = 10$ [%] ($\Delta_h = 5$ [%], $\Delta_t = 5$ [%])
Short spreading codes	Walsh Hadamard codes with $K^{MC} = 32$

Table 9.4
Simulation Parameters for DS-CDMA

Short spreading codes	Gold codes with $K^{DS} = 31$

Figure 9.10, as the number of Rake fingers increases, the BER performance becomes better. For the channel with 18 paths, the Rake combiner with 18 fingers could perform best, but it would be impossible to implement. Comparing Figures 9.9 and 9.10 reveals that, in the downlink, as compared with the DS-CDMA system with the 18-finger Rake combiner, the MC-CDMA system with the MMSE performs better for a large number of users, although it performs worse or is comparable up to the region of a middle number of users because of the energy loss in the guard interval. Compared with the DS-CDMA system with the 5-finger Rake combiner, it always performs better.

On the other hand, Figures 9.11 and 9.12 show the BER performance in the uplink for the MC-CDMA and DS-CDMA systems, respectively. In Figure 9.11, the performance of the decorrelating multiuser detection scheme is very poor. This is because the degradation due to the noise enhancement is dominant even if the code orthogonality among users is restored. The performance of the MMSE multiuser detection scheme is excellent, and it can keep a good BER for a larger number of users. However, as long as we employ the guard interval, we cannot avoid some degradation from the lower bound. On the other hand, for the DS-CDMA system, Figure 9.12 shows

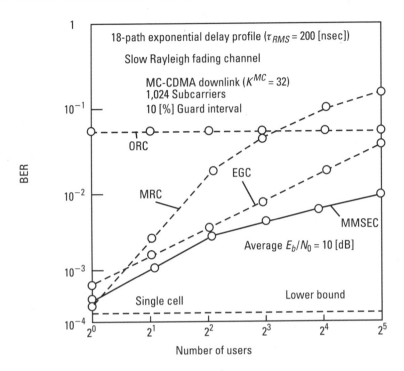

Figure 9.9 BER performance of an MC-CDMA system in downlink.

the effectiveness of the SIC scheme. The performance of the 18-finger SIC scheme is excellent, but it also would be impossible to implement. Finally, similar to the conclusion in the downlink, a comparison of Figures 9.11 and 9.12 reveals that, in the uplink, as compared with the DS-CDMA system with the SIC scheme, the MC-CDMA system with the MMSE multiuser detection scheme is comparable up to the region of a middle number of users and performs better in the region of a larger number of users.

9.4.7   Sliding DFT-Based Subcarrier Recovery Method

Let us focus our attention on a subcarrier recovery problem for the MMSEC scheme in a downlink. Figure 9.13 shows a signal burst format, where the preamble is composed of three *discontinuous* known pilot symbols to perform subcarrier recovery and some data symbols to demodulate. In the preamble part, no signals are multiplexed and we can freely design the format. Therefore, assuming

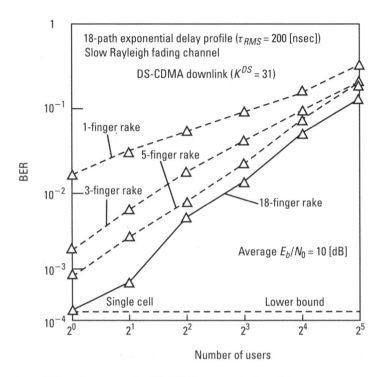

Figure 9.10 BER performance of a DS-CDMA system in downlink.

$$d_{j'}(m) = \frac{1}{\sqrt{K^{MC}}}, \ (m = 0, 1, \ldots, K^{MC} - 1) \qquad (9.71)$$

$$c_{j'}(Pm + p' + iK^{MC}) = \frac{1}{\sqrt{K^{MC}}}, \ (m = 0, 1, \ldots, K^{MC} - 1) \qquad (9.72)$$

we can rewrite (9.35) and (9.36) in vector forms as (we drop the subscripts for the user index, j and j')

$$D_{p'}^{MC}(i) = \mathbf{G}_{p'}^{H}(i)\mathbf{y}_{p'}(i) \qquad (9.73)$$

$$\mathbf{G}_{p'}(i) = [G_{p',i}(0), \ldots, G_{p',i}(K^{MC} - 1)]^{T} \qquad (9.74)$$

$$\mathbf{y}_{p'}(i) = [y_{p',i}(0), \ldots, y_{p',i}(K^{MC} - 1)]^{T} \qquad (9.75)$$

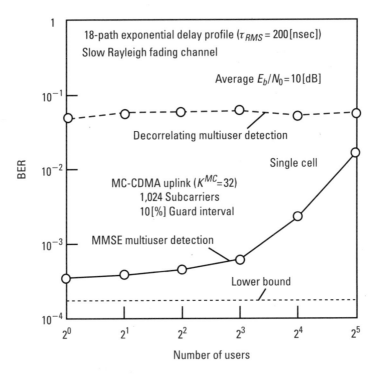

Figure 9.11 BER performance of an MC-CDMA system in uplink.

where $\mathbf{G}_{p'}(i)$ is the $(K^{MC} \times 1)$ gain vector to determine and $\mathbf{y}_{p'}(i)$ the $(K^{MC} \times 1)$ baseband component vector for the received pilot symbol. In this case, (9.41) can be simplified to

$$\text{minimize MSE}(\mathbf{G}_{p'}(i)) = E\left[\left| a_{p'}(i) - \mathbf{G}_{p'}^{H}(i)\mathbf{y}_{p'}(i) \right|^2 \right] \quad (9.76)$$

We can solve the above minimization problem with any adaptive algorithm. For instance, if we use the normalized least mean square (LMS) algorithm, we have

$$e(i) = a_{p'}(i) - \mathbf{G}_{p'}^{H}(i)\mathbf{y}_{p'}(i) \quad (9.77)$$

$$\mathbf{G}_{p'}(i+1) = \mathbf{G}_{p'}(i) + \frac{\mu}{\epsilon + \left| \mathbf{y}_{p'}(i) \right|^2} \mathbf{y}_{p'}(i)e*(i) \quad (9.78)$$

where ϵ is a positive constant and μ is an adaptation constant to control the convergence.

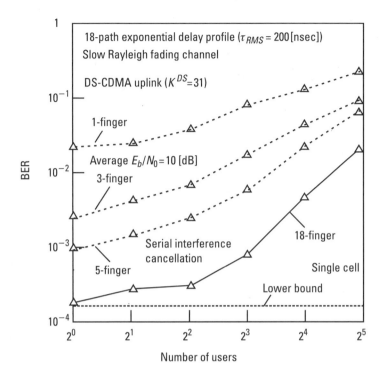

Figure 9.12 BER performance of a DS-CDMA system in uplink.

The adaptive subcarrier recovery method looks smart but suffers from the very small number of possible iterations. This is because $\mathbf{y}_{p'}(i)$ is calculated from the DFT of the received pilot symbol, so only one vector can be obtained from each pilot symbol, although it is composed of a lot of samples. For instance, in Figure 9.13(a), the preamble is composed of the three *discontinuous* pilot symbols, so we can get only three iterations for the adaptive algorithm, and this is insufficient to get a good convergence.

Figure 9.13(b) shows a sliding DFT-based adaptive subcarrier recovery method that can increase the number of iterations [17, 18]. Similar to the DFT-based method shown in Figure 9.13(a), the preamble is composed of three known pilot symbols, but they are continuously connected. Therefore, we can have many iterations by sliding the DFT with sliding width of δ_s. In the ith iteration ($i = 1, 2, \ldots$), the sliding DFT gives a phase rotation $\theta = -2\pi m P \delta_s (i - 1)/t_s$ for the mth component of $\mathbf{y}_{p'}(i)$, we need to compensate for the phase rotation. The adaptive algorithm is given by

$$e(i) = a_{p'}(i) - \mathbf{G}_{p'}^{H}(i)\,\mathbf{\Theta}(i)\,\mathbf{y}_{p'}(i) \tag{9.79}$$

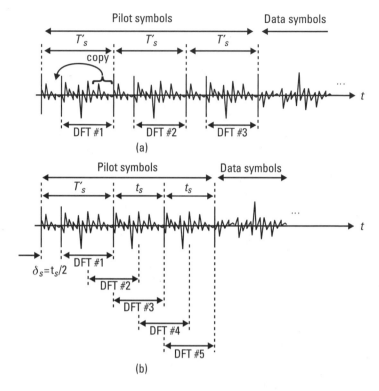

Figure 9.13 Subcarrier recovery method for MMSEC: (a) subcarrier recovery method using DFT; and (b) subcarrier recovery method using sliding DFT.

$$\mathbf{G}_{p'}(i+1) = \mathbf{G}_{p'}(i) + \frac{\mu}{\epsilon + |\mathbf{y}_{p'}(i)|^2} \mathbf{\Theta}_{p'}(i)\mathbf{y}_{p'}(i)e*(i) \quad (9.80)$$

In (9.80), $\mathbf{\Theta}_{p'}(i)$ is the $(K^{MC} \times K^{MC})$ phase compensation matrix given by

$$\mathbf{\Theta}_{p'}(i) = diag\left[e^{j2\pi\frac{0P+p'}{t_s}\delta_s(i-1)}, \ldots, e^{j2\pi\frac{(K^{MC}-1)P+p'}{t_s}\delta_s(i-1)} \right] \quad (9.81)$$

where $diag[\ldots]$ is the diagonal matrix.

Table 9.5 shows the simulation parameters to evaluate the BER performance of the sliding DFT-based MMSEC, and Figure 9.14 shows the obtained results. Here, we assume $\delta_s = t_s/2$, so we can have seven iterations for the normalized LMS algorithm. This figure also compares the performance

Table 9.5
Simulation Parameters for Evaluation of the Sliding DFT-Based MMSEC

Information transmission rate	4 [Mbps] (QPSK)
Short spreading codes	Walsh Hadamard codes with $K^{MC} = 32$
Number subcarriers	512
Combining scheme	MMSEC
Preamble	4 [pilot symbols]
Payload	60 [data symbols]
Number of users	3
Adaptive algorithm	Normalized LMS with $\mu = 1.0$ and $\epsilon = 1.0$ (optimized)
Channel fading	Frequency selective fast Rayleigh
Multipath delay profile	10-path exponentially decaying
RMS delay spread	200 [nsec]
Doppler frequency	80 [Hz]

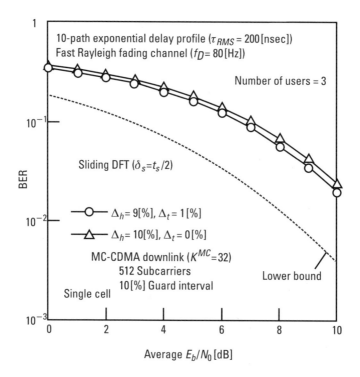

Figure 9.14 BER performance of sliding DFT-based subcarrier recovery method.

between the conventional head guard interval insertion method and the head/ tail guard interval insertion method for the same guard interval efficiency $\Delta_G = 10\%$. The performance with the head/tail guard interval is superior to that only with the head guard interval, but even using the sliding DFT-based method, there is still a large difference in BER between the adaptive MMSEC and the lower bound.

9.5 Conclusions

We showed that the MC-CDMA system is well suited for a high-speed data transmission, whereas the DS-CDMA system could see more than around 10 paths in the multipath delay profile. A DS-CDMA receiver would lose almost all of the received signal energy, whereas a MC-CDMA would effectively collect it, although a fraction of the energy would be lost in the guard interval.

We believe that the MC-CDMA system must be one of the major candidates for 4G mobile communications where high scalability is required in possible transmission rates. Intensive research continues to be conducted on the MC-CDMA system. However, to make it a real candidate, much more research is required.

References

[1]   Rhee, M. Y., *CDMA Cellular Mobile Communications and Network Security*, Upper Saddle River, NJ: Prentice Hall, 1998.

[2]   Prasad, R., *CDMA for Wireless Personal Communications*, Norwood, MA: Artech House, 1996.

[3]   Ojanpera, T., and R. Prasad (eds.), *Wideband CDMA for Third Generation Mobile Communications*, Norwood, MA: Artech House, 1998.

[4]   Yee, N., J. P. Linnartz, and G. Fettweis, "Multicarrier CDMA in Indoor Wireless Radio Networks," *Proc. of IEEE PIMRC'93*, Yokohama, Japan, September 1993, pp. 109–113.

[5]   Fazel, K., and L. Papke, "On the Performance of Convolutionally-Coded CDMA/ OFDM for Mobile Communication System," *Proc. of IEEE PIMRC'93*, Yokohama, Japan, September 1993, pp. 468–472.

[6]   Chouly, A., A. Brajal, and S. Jourdan, "Orthogonal Multicarrier Techniques Applied to Direct Sequence Spread Spectrum CDMA Systems," *Proc. of IEEE GLOBECOM'93*, Houston, TX, November 1993, pp. 1723–1728.

[7]   Hara, S., T. H. Lee, and R. Prasad, "BER Comparison of DS-CDMA and MC-CDMA for Frequency Selective Fading Channels," *Proc. of the 7th International*

Thyrrhenian Workshop on Digital Communications, Viareggio, Italy, September 10–14, 1995, pp. 3–14.

[8] Prasad, R., and S. Hara, "CDMA-Based Hybrid Multiple Access Schemes for Wireless Multimedia Communications," *Technical Document of European Cooperation in the Field of Scientific and Technical Research (COST) 231,* Belfort, France, January 24–26, 1996, pp. TD(96)020.1–TD(96)020.16.

[9] Hara, S., and R. Prasad, "DS-CDMA, MC-CDMA, and MT-CDMA for Mobile Multi-Media Communications," *Proc. of the 46th IEEE VTC,* Atlanta, GA, April 18–May 1, 1996, pp. 1106–1110.

[10] Prasad, R., and S. Hara, "An Overview of Multicarrier CDMA," *Proc. of the 4th IEEE International Symposium on Spread Spectrum Techniques and Applications (ISSSTA'96),* Mainz, Germany, September 22–25, 1996, pp. 107–114.

[11] Hara, S., and R. Prasad, "Overview of Multicarrier CDMA," *IEEE Communications Magazine,* Vol. 35, No. 12, December 1997, pp. 126–133.

[12] Kleer, F., S. Hara, and R. Prasad, "Performance Evaluation of a Successive Interference Cancellation Scheme in a Quasi-Synchronous MC-CDMA System," *Proc. of IEEE ICC'98,* Atlanta, GA, June 7–11, 1998, pp. 370–374.

[13] Hara, S., and M. Budsabathon, "Spread Spectrum-Based Subcarrier Recovery Method for Multicarrier Code Division Multiplexing System," *European Transactions on Tele-communications,* Vol. 10, No. 4, July/August 1999, pp. 369–376.

[14] Hara, S., and R. Prasad, "Design and Performance of Multicarrier CDMA System in Frequency Selective Fading Channels," *IEEE Trans. on Veh. Technol.,* Vol. 48, No. 9, September 1999, pp. 1584–1595.

[15] Hara, S., "Multicarrier CDMA—A Promising Transmission and Multiple Access Technique for Fourth-Generation Mobile Communications Systems," *Proc. of URSI International Symposium on Signals, Systems, and Electronics (ISSSE)'01,* Tokyo, Japan, July 24–27, 2001, pp. 238–241.

[16] Tsumura, S., and S. Hara, "Design and Performance of Quasi-Synchronous Multicarrier CDMA Uplink," *Proc. of IEEE VTC 2001-Fall,* Atlantic City, NJ, October 7–11, 2001, pp. 843–847 (available in CD-ROM) .

[17] Tsumura, S., and S. Hara, "MMSE-Based Adaptive Equalizer with Effective Use of Pilot Signal for MC-CDM System," *Proc. of IEEE ISSSTA2002,* Prague, Czech, September 2002.

[18] Tsumura, S., and S. Hara, "A Novel Subcarrier Recovery Method for Multicarrier CDM System," *Proc. of IEEE VTC2002-Fall,* Vancouver, Canada, September 2002.

[19] Monsen, P., "Digital Transmission Performance on Fading Dispersive Diversity Channels," *IEEE Trans. Commun.,* Vol. COM-21, January 1973, pp. 33–39.

10

Future Research Directions

10.1 Introduction

Recently, there have been several sessions on Beyond 3G or 4G Systems at major international conferences on wireless communications, where different speakers have given different images and concepts for 4G mobile communications systems. We often asked prominent researchers at these conference venues the question, "What are 4G systems?" but even their answers have been very different. For instance, some systems are based on multiple access techniques in multiple cell environments, while others are based on random access techniques in isolated cell environments. In addition, some services are provided in microwave frequency bands, while others are provided in millimetric frequency bands. This can be very confusing. We still might not know the answer to our question, or perhaps all the answers are correct. Such ambiguity is why it is worthwhile for academic and company researchers to continue to conduct research on 4G systems. Instead of providing a conclusion for the book, this chapter offers some interesting research topics related to multicarrier techniques for future research directions.

Section 10.2 discusses two possible paths toward the realization of OFDM-based 4G systems, such as a migration from cellular phone systems and a migration from wireless LANs. Section 10.3 introduces two variants based on the MC-CDMA scheme, including orthogonal frequency and code division multiplexing (OFCDM). Section 10.4 shows the application of an adaptive array antenna for the OFDM scheme and discusses four different kinds of array configurations. Section 10.5 introduces a multiple input and multiple output (MIMO) system and discusses the suitability of OFDM signaling for the MIMO system. Section 10.7 shows a linear amplification method of the OFDM signal with two nonlinear power amplifiers. Finally, Section 10.8 outlines future research topics.

10.2 Where Will 4G Systems Come From?

Figure 10.1 shows two possible paths toward the realization of 4G systems, such as a migration from cellular phone systems and a migration from wireless LANs.

Cellular phone systems have never experienced more than 30-Mbps data transmission even in the current 3G [1] and 3.5G [2] standards; therefore, as a totally new scheme, the OFCDM scheme [3] is proposed for 4G downlink. The OFCDM scheme is really a variant based on MC-CDM and is suited for service provision in cellular or multiple cell environments like the DS-CDMA scheme in 3G and 3.5G systems.

On the other hand, as shown in Section 8.3, wireless LANs can already provide up to 54-Mbps data transmission even in the current standards, such as IEEE802.11a [4], HIPERLAN/2 [5], and MMAC [6, 7], although the service provision is still limited for stationary or low mobility users. Therefore, if a future wireless LAN standard, which might be similar to the current OFDM-based standards with a bit higher transmission rate, can cope with high mobility of users, we will be able to call it a 4G system.

10.3 Variants Based on MC-CDMA Scheme

10.3.1 OFCDM System

Section 9.4 discussed the details of the conventional MC-CDMA system, where we ignored the interleaver in Figure 9.3(a) and the corresponding deinterleaver in Figure 9.3(c).

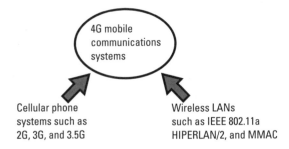

Figure 10.1 Two possible paths toward 4G systems.

When employing the interleaver with interleaving depth of $P(\times K^{MC})$ in Figure 9.3(a), the power spectrum of the resultant signal changes into Figure 10.2. No change appears in the whole shape of the spectrum but spread data are mapped onto successive K^{MC} subcarriers. The power spectrum has distinct P blocks, each of which is composed of K^{MC} subcarriers and conveys information for an individual data. This modified MC-CDMA system is proposed for downlink and called an "OFCDM system" [3]. Figures 10.3 and Figure 10.4 compare the concept of a conventional MC-CDM(A) system and an OFCDM system.

The advantage of an OFCDM system is its own high scalability in possible transmission rates, namely, it can support low-to-high information transmission rates by employing code, time, and frequency division multiplexing. Figure 10.5 shows a code and time division multiplexing in the

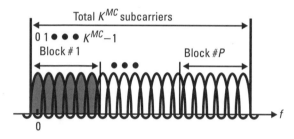

Figure 10.2 Power spectrum of an OFCDM signal.

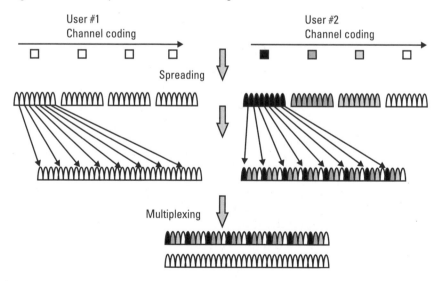

Figure 10.3 Concept of a conventional MC-CDM(A) system.

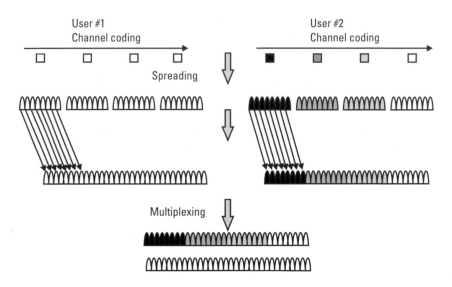

Figure 10.4 Concept of an OFCDM system.

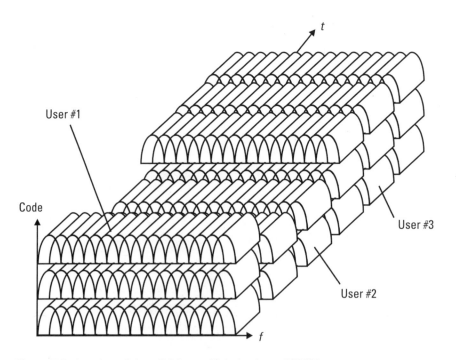

Figure 10.5 A code and time division multiplexing in an OFCDM system.

OFCDM system, where different numbers of codes and time slots are flexibly allocated for users. Figure 10.6 shows a frequency division multiplexing in the OFCDM system. Different subcarriers are allocated for different users. The spread data for an individual user are transmitted over the blocked subcarriers, so the OFCDM system could reduce the receiver hardware complexity, because low-rate users do not have to receive all the subcarriers, and only a fraction of the spectrum. The disadvantage might be that the OFCDM system cannot obtain full diversity effect, because the spread operation is done within a block and the subcarriers in a block would be highly correlated. However, the use of channel coding before spreading also means spreading information over different blocks, so it gives an almost full diversity effect, and the code spreading over the correlated subcarriers mitigates the distortion of the orthogonality.

The OFCDM system furthermore changes the spreading factor (SF) according to cell layouts [3]. If we take a code division multiplexing approach in wireless cellular environments, frequency selective fading distorts the orthogonality among the spreading codes but we can mitigate intercell interference by means of the long spreading codes. Therefore, in a multiple cell environment where intercell interference is dominant, the variable spreading factor (VSF)-OFCDM system increases *SF*, whereas in an isolated cell environment where there is no intercell interference, it decreases *SF* to unity

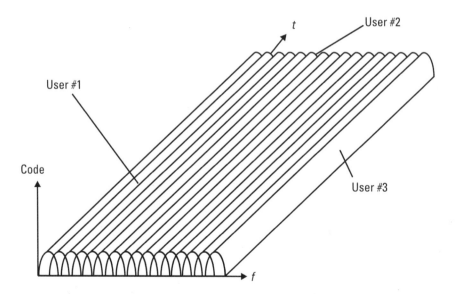

Figure 10.6 A frequency division multiplexing in an OFCDM system.

because the OFDM ($SF = 1$) system suffers from no MUI. Figure 10.7 shows the concept of the VSF-OFCDM system.

Intensive research related to the VSF-OFCDM system has been done so far, including the performance comparison among OFDM-TDMA, DS-CDMA, multicarrier/DS-CDMA, and MC-CDMA systems in multiple cell environments [8, 9], the performance comparison between VSF-OFCDM and OFDM-TDM forward link in multiple cell environments [3], and the proposal of a three-step cell search for MC-CDMA forward link [10].

The OFCDM system, which spreads the information sequence in the frequency domain, can achieve a better BER performance for lower-level modulation such as BPSK and QPSK [3]. However, when employing higher-level modulations such as 16 QAM at subcarriers, the attainable BER performance becomes poor. This is because the frequency domain spreading cannot compensate well for the severe distortion of the code orthogonality due to frequency selective fading [11]. An OFCDM system with not only frequency domain spreading but also time domain spreading is proposed to mitigate the distortion [11], and the concept is extended to a VSF-OFCDM system [12]. Table 10.1 shows typical transmission parameters discussed for the OFCDM system.

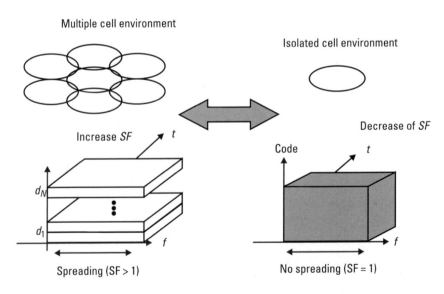

Figure 10.7 Change of the spreading factor according to cell layouts.

Table 10.1
Transmission Parameters for OFCDM System

Bandwidth	101.5 [MHz]
Number of IDFT/DFT points	1,024
Number of subcarriers	768
Subcarrier separation	131.836 [kHz]
Useful symbol length	7.585 [μsec]
Guard interval length	1.674 [μsec]
Short spreading codes	Walsh Hadamard with $K^{MC} = 1$-to-32
Long spreading codes	Pseudo random
Modulation	QPSK, 16 QAM
Combining scheme	MMSEC
Preamble	4 [pilot symbols]
Payload	48-50 [data symbols]
Channel coding/decoding	Turbo coding ($R = 1/2$, $K = 4$)
	Max-Log-MAP decoding (8 iterations)

10.3.2 Other Variant Based on MC-CDMA Scheme

Shannon's water filling theorem tells us how the power should be allocated over the channels when parallel channels are given with a power constraint. Namely, if the transmit side knows the channel condition, it should allocate more powers to channels with lower noise powers. Therefore, in the case of multicarrier transmission, if the transmitter knows the instantaneous frequency response of the channel, it should allocate more powers to subcarriers with higher gains; in other words, for equal power allocation, it should allocate information bits so as to make the SNR per bit as equal as possible over different subcarriers (bit loading).

In 2002, an MC-CDM system based on the Shannon's Water Filling Theorem was proposed for a downlink. It is called "MC-CDM system with frequency scheduling" [13, 14]. Note that, unlike the case of a point-to-point data transmission, a multiplexing system needs to take into consideration the inter- and intracell interference power, as well as the noise power, because it suffers from multiple access interference and intercell interference.

Figure 10.8 shows the concept of an MC-CDM system with frequency scheduling. First, a base station transmits a common pilot signal to all users [see Figure 10.8(a)], and each mobile user estimates its received signal to noise plus interference power ratio (SNIR) in block by block [see Figure 10.8(b)]. Then, each user notifies the signal qualities (SNIRs) to the base station in a piggyback manner [see Figure 10.8(c)]. Finally, based on the

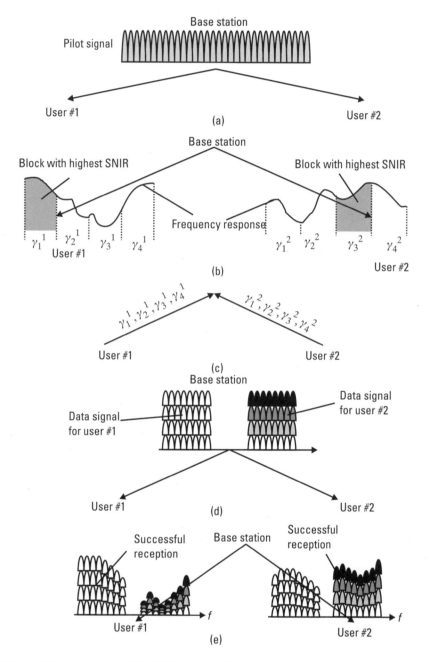

Figure 10.8 Concept of an MC-CDM system with frequency scheduling: (a) pilot signal transmission; (b) block-by-block SNIR estimation; (c) SNIR (block quality) notification; (d) data signal transmission; and (e) data signal reception.

SNIR table obtained from all the users, the base station determines which blocks should be used for an individual user [see Figure 10.8(d)]. The data transmission is made only in blocks with higher SNIRs for each user, so the frequency scheduling scheme can much improve the BER performance and enhance the system performance [see Figure 10.8(e)].

10.4 OFDM Adaptive Array Antennas

10.4.1 Principle of Adaptive Array Antenna

An adaptive array antenna is an antenna that controls its own pattern, by means of feed-back or feed-forward control [15, 16]. Here, we consider only receiving array antennas.

Figure 10.9 shows the basic configuration of an adaptive array antenna and its antenna beam pattern. An adaptive array antenna changes its antenna pattern through optimization of the SNIR, by changing the values of array weights [see Figure 10.9(a)]. It does not need to know the arrival direction of interference or desired signal in advance, because an antenna pattern can automatically reject the interference direction by null and track the desired signal direction [see Figure 10.9(b)].

So far, adaptive array antennas have been discussed for the suppression of cochannel interferers, delayed signals beyond the guard interval, and Doppler shifted signals in an OFDM scheme [17–26]. As shown in Sections 10.4.2 and 10.4.3, there are four different types of adaptive array antennas considered applicable for the OFDM scheme.

10.4.2 Post-FFT and Pre-FFT Type OFDM Adaptive Array Antennas

Figure 10.10 shows two types of adaptive array antennas applicable for the OFDM scheme. Figure 10.10(a) shows a post-FFT type OFDM adaptive array antenna [17], where there is one OFDM demodulator, including the FFT at each antenna element. Using this configuration, the weighted outputs are combined at each subcarrier, so it requires high computational complexity, although the attainable performance would be better.

On the other hand, Figure 10.10(b) shows a pre-FFT type OFDM adaptive array antenna [18–26], where there is only one OFDM demodulator. Using this configuration, the weighted array outputs are combined just before the OFDM demodulator, so it can reduce computational complexity, although the attainable performance would be inferior to that of the post-FFT type adaptive array antenna.

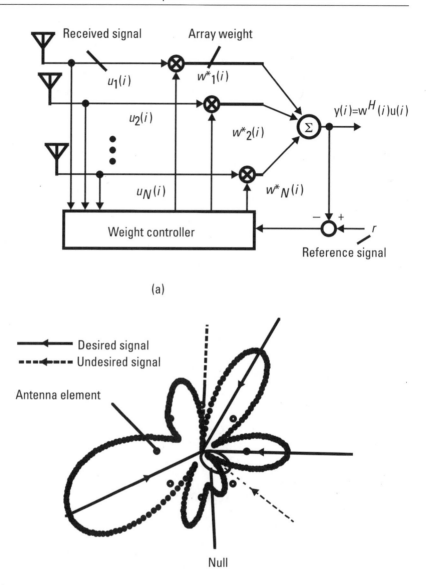

(a)

(b)

Figure 10.9 Principle of an adaptive array antenna: (a) basic configuration; and (b) antenna beam pattern.

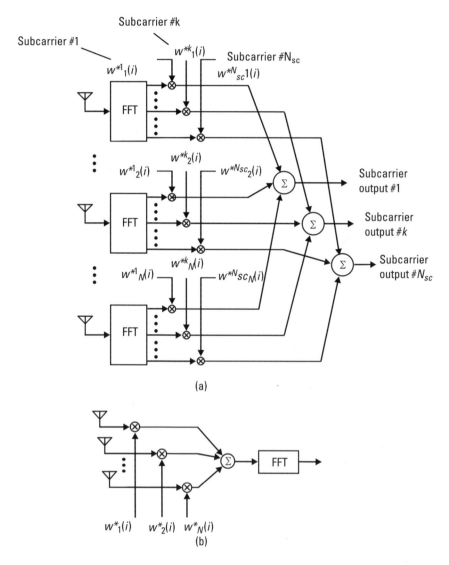

Figure 10.10 Two types of adaptive array antennas: (a) a post-FFT type; and (b) a pre-FFT type.

10.4.3 Weight-Per-User and Weight-Per-Path Type OFDM Adaptive Array Antennas

Figure 10.11 shows two types of adaptive array antennas, which are workable in multipath environments. Figure 10.11(a) shows a weight-per-user type

Received signals from one user

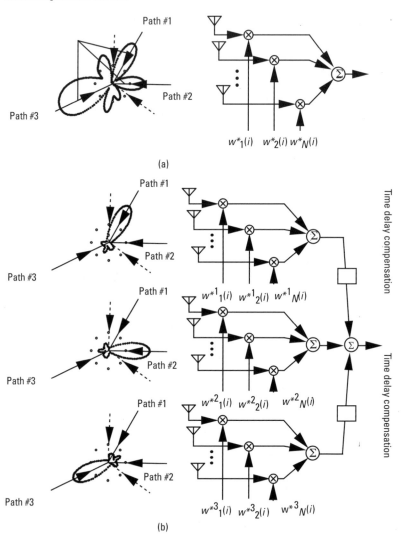

Figure 10.11 Two types of adaptive array antennas: (a) a weight-per-user type; and (b) a weight-per-path type.

adaptive array antenna [17–26], which controls its own antenna beam pattern for incoming signals from a desired user with a set of array weights.

On the other hand, Figure 10.11(b) shows a weight-per-path type adaptive array antenna, which controls its own antenna beam pattern for each incoming signal through a different path from the same desired user with a set of array weights. It is very clear that the weight-per-user type array antenna is much simpler than the weight per path-type array antenna. Note that the weight-per-path type array antennas are applicable for not only multicarrier transmission but also singlecarrier transmission, whereas the weight-per-user type arrays are applicable only for multicarrier transmission, including OFDM and SS-based transmission. This is because the OFDM and SS signals are inherently robust to multipath fading.

10.5 MIMO-OFDM

When a transmitter and a receiver, with an appropriate channel coding/decoding scheme, are equipped with multiple antennas, the presence of multipath fading can improve achievable transmission rates [27]. For such MIMO channels, several optimum space-time codes have been designed, assuming that the transmitter does not know the structure of the channel [28, 29].

Figure 10.12 shows a MIMO system, where the transmitter has M-element antennas and the receiver has N-element antennas. For the MIMO channel, we can define the channel matrix $(M \times N)$ as

$$\mathbf{H} = \{h_{ij}\} \qquad (10.1)$$

where h_{ij} means the path gain between the ith transmit antenna element and the jth receive antenna element. In general, the MIMO system requires a flat fading characteristic at each subchannel, in other words, h_{ij} should be

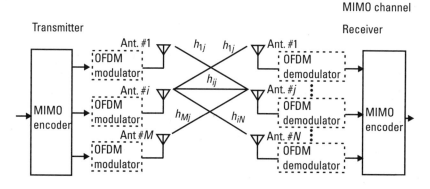

Figure 10.12 A MIMO system.

a complex value. Therefore, if the transmission rate is high enough to make the whole channel frequency selective, it requires an OFDM signaling to have a flat fading over each subchannel. In this sense, the OFDM scheme is suited for the MIMO system [30]. Field trial results on a 4G MIMO-OFDM system have been reported [31].

As shown, the design of space-time codes assumes no knowledge on the structure of the transmission channel at the transmit side. If the transmitter knows the channel structure, a joint transmitter/receiver optimization is furthermore possible for the multiple transmit/receive antennas scenario [32, 33]. The OFDM scheme can play an important role in such a situation.

10.6   Linear Amplification of OFDM Signal with Nonlinear Components

As shown in Section 4.9, OFDM signals are much more sensitive to non-linear amplification than singlecarrier-modulated signals. Therefore, when amplifying the OFDM signals with a nonlinear power amplifier, a larger input back-off is required to reduce the spectrum spreading, and it results in a low-power efficiency.

Nonlinear amplification has been considered to be unavoidable. Indeed, many techniques to reduce the spectrum spreading have been proposed for OFDM systems, such as deliberately clipping [34] and so on. However, almost all the techniques require complicated signal processing or show some performance degradation, so there has been no "killer" technique to make the OFDM signals robust against nonlinear amplification.

In 2002, a semiconductor company was successful in a linear amplification of OFDM signals with nonlinear amplifiers [35], which was based on the linear amplification with nonlinear components (LINC) method.

The LINC is an old technique [36], which dates back to 1935 [37], and is based on a fact that any bandpass signal can be represented by two constant-envelope phase-modulated signals.

Figure 10.13 shows the principle of the LINC method. Define an input signal as

$$V_{in}(t) = a(t) e^{j\Theta(t)} \tag{10.2}$$

where $a(t)$ is an envelope and $\Theta(t)$ is a phase. The input signal $V_{in}(t)$ can be represented by the following two signals, $S_1(t)$ and $S_2(t)$, with a constant envelope of $V_m \left(\geq |a(t)| \right)$:

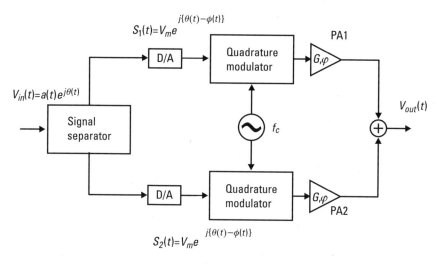

Figure 10.13 Principle of the LINC method.

$$V_{in}(t) = S_1(t) + S_2(t) \qquad (10.3)$$

$$S_1(t) = V_m e^{j\{\theta(t)+\phi(t)\}} \qquad (10.4)$$

$$S_2(t) = V_m e^{j\{\theta(t)-\phi(t)\}} \qquad (10.5)$$

Substituting (10.4) and (10.5) into (10.3) leads to

$$\phi(t) = \cos^{-1}\left\{\frac{a(t)}{2V_m}\right\} \qquad (10.6)$$

Figure 10.14 shows a LINC signal separation. Even if an input signal has an arbitrary envelope and an arbitrary phase, there always exist two signals with the same constant envelope, the sum of which is equal to the input signal.

$S_1(t)$ and $S_2(t)$ have the constant envelope, so they can be amplified with highly power-efficient and highly nonlinear power amplifiers. Defining the gain and phase of the two nonlinear amplifiers as G and φ, the output signal is given by

$$V_{out}(t) = G e^{j\varphi}\{S_1(t)e^{j2\pi f_c t} + S_2(t)e^{j2\pi f_c t}\} \qquad (10.7)$$

$$= G V_{in}(t) e^{j(2\pi f_c t+\varphi)}$$

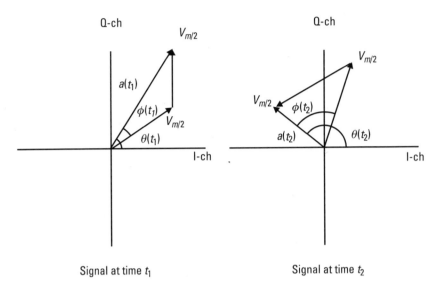

Signal at time t_1 Signal at time t_2

Figure 10.14 A LINC signal separation.

The LINC method can work quite well if the characteristics (AM/ AM, AM/PM, PM/AM and PM/PM conversion characteristics) of the two nonlinear amplifiers are all the same. However, it is practically impossible. If there is a mismatch between them, the LINC method not only cannot reduce the spectrum spreading but also generates some inband distortion in the amplified OFDM signals.

Figure 10.15 shows a combination of a predistortion and the LINC to compensate for the gain/phase mismatch. In the figure, e is defined as the error between the input signal and the amplified output. The predistorter gives the input signal a predistortion so as to minimize the error. Several mismatch cancellation methods have been proposed [38, 39].

10.7 Conclusions

To show future research directions, we briefly presented several recent interesting research topics related to multicarrier techniques. We believe that multicarrier techniques will play important roles in 4G systems, however, to make the multicarrier a core physical layer technique in 4G systems, there are a lot of future research areas we should investigate further.

For instance, there has been intensive research recently all over the world on variants based on the MC-CDMA scheme. However, few works

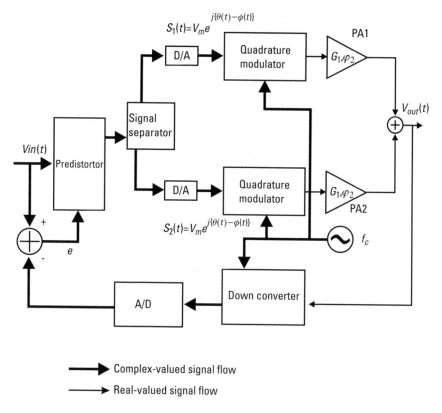

$S_1(t)=V_m e^{j\{\theta(t)-\phi(t)\}}$

Figure 10.15 A combination of a predistortion and the LINC.

have been dedicated to pure OFDM-based schemes aimed at 4G systems, with emphasis on the signal format to cope with the Doppler shift due to mobile motions. We think that one way to give the system a robustness against the Doppler shift is through the use of scattered pilots (see Section 8.2).

In terms of access protocols, no one knows whether CDMA is really suited for the specification in 4G systems. OFDM-TDMA and OFDM-CSMA/CA, as well as MC-CDMA systems, are all candidates. The performance comparison of these systems in multiple and isolated cell environments will be required.

Adaptive array antennas can enhance the transmission performance for OFDM-based systems. As shown in Section 10.4, there are many different ways to configure array antennas and OFDM demodulators. This has been a recent hot topic on 4G systems, however, further investigation, taking into consideration the trade-off between the receiver complexity and the attainable performance, will be required.

For MIMO-OFDM, this has also been a recent hot topic in wireless communications in conjunction with adaptive array and diversity antennas. However, we have never seen the capacity analysis of a MIMO-OFDM system that can jointly suppress cochannel interference from other cells.

Finally, for the LINC method, there are a lot of ways for gain/phase mismatch cancellation. Examining the trade-off between hardware complexity and attainable performance will be important.

References

[1]   Ojanpera, T., and R. Prasad (eds), *Wideband CDMA for Third Generation Mobile Communications,* Norwood, MA: Artech House, 1998.

[2]   3 GPP, 3G TR25.848, V.0.6.0, May 2000.

[3]   Atarashi, H., and M. Sawahashi, "Variable Spreading Factor Orthogonal Frequency and Code Division Multiplexing (VSF-OFCDM)," *Proc. of 2001 Third International Workshop on Multicarrier Spread-Spectrum and Related Topics (MC-SS2001),* September 2001, pp. 113–122.

[4]   IEEE Std. 802.11a, "Wireless Medium Access Control (MAC) and Physical Layer (PHY) Specifications: High-Speed Physical Layer Extension in the 5-GHz Band," IEEE, 1999.

[5]   ETSI TR 101 475, "Broadband Radio Access Networks (BRAN); HIPERLAN Type 2; Physical (PHY) Layer," ETSI BRAN, 2000.

[6]   ARIB STD-T70, "Lower Power Data Communication Systems Broadband Mobile Access Communication System (CSMA)," ARIB, December 2000.

[7]   ARIB STD-T70, "Lower Power Data Communication Systems Broadband Mobile Access Communication System (HiSWANa)," ARIB, December 2000.

[8]   Abeta, S., et al., "Forward Link Capacity of Coherent Multicarrier/DS-CDMA and MC-CDMA Broadband Packet Wireless Access in a Multicell Environment," *Proc. of IEEE VTC2000-Fall,* Boston, MA, September 2000, pp. 2213–2218.

[9]   Atarashi. H., S. Abeta, and M. Sawahashi, "Broadband Packet Wireless Access Appropriate for High-speed and High-Capacity Throughput," *Proc. of IEEE VTC2001-Spring,* Rhodes, Greece, May 2001, pp. 566–570.

[10] Hanada, Y., K. Higuchi, and M. Sawahashi, "Three-Step Cell Search Algorithm for Broadband Multicarrier CDMA Packet Wireless Access," *Proc. of IEEE PIMRC'01,* September 2001.

[11] Miyoshi, K., A. Matsumoto, and M. Uesugi, "A Study on Time Domain Spreading for OFCDM (in Japanese)," *IEICE Technical Report,* RCS2001-179, November 2001, pp. 13–18.

[12] Kishiyama, Y., et al., "VSF-OFCDM withTwo-Dimensional Spreading Prioritizing Time Domain Spreading in Other-Cell Interference Environment (in Japanese)," *IEICE Technical Report,* RCS2002-133, July 2002, pp. 87–92.

[13] Hara, Y., et al., "MC-CDM System for Packet Communications Using Frequency Scheduling (in Japanese)," *IEICE Technical Report,* RCS2002-129, July 2002, pp. 61–66.

[14] Hara, Y., et al., "Frame Configuration and Control Scheme in MC-CDM Systems with Frequency Scheduling (in Japanese)," *IEICE Technical Report,* RCS2002-130, July 2002, pp. 67–72.

[15] Compton, Jr., R. T., *Adaptive Antennas Concepts and Performance,* Englewood Cliffs, NJ: Prentice Hall, 1988.

[16] Monzingo, R. A., and T. W. Miller, *Introduction to Adaptive Arrays,* New York: John Wiley & Sons, 1980.

[17] Li, Y., and N. R. Sollenberger, "Adaptive Antenna Arrays for OFDM Systems with Cochannel Interference," *IEEE Trans. Commun.,* Vol. 47, No. 2, February 1999, pp. 217–229.

[18] Nishikawa, A., Y. Hara, and S. Hara, "OFDM Adaptive Array for Doppler-Shifted Wave Suppression," *Technical Report of IEICE,* RCS-2000-113, October 2000, pp. 57–62.

[19] Nishikawa, A., Y. Hara, and S. Hara, "A Study on OFDM Adaptive Array in Mobile Communications," *Technical Report of IEICE,* RCS-2000-232, March 2001, pp. 73–78.

[20] Nishikawa, A., Y. Hara, and S. Hara, "An OFDM Adaptive Array for Doppler-Shifted Wave Suppression," *Proc. 2001 IEICE General Conference,* B-5-151, March 2001, p. 549.

[21] Hara, S., A. Nishikawa, and Y. Hara, "A Novel OFDM Adaptive Antenna Array for Delayed Signal and Doppler-Shifted Signal Suppression," *Proc. IEEE International Conference on Communications (ICC) 2001,* Helsinki, Finland, June 11–14, 2001, pp. 2302–2306.

[22] Hara, S., S. Hane, and Y. Hara, "Adaptive Antenna Array for Reliable OFDM Transmission," *Proc. 6th International OFDM Workshop,* Hamburg, Germany, September 18–19, 2001, pp. 1.1–1.4.

[23] Hane, S., Y. Hara, and S. Hara, "Selective Signal Reception for OFDM Adaptive Array Antenna," *Technical Report of IEICE,* AP-2001-69, August 2001, pp. 35–41.

[24] Hane, S., S. Hane, and Y. Hara, "Does OFDM Really Prefer Frequency Selective Fading Channels," *Proc. 2001 Third International Workshop on Multicarrier Spread-Spectrum (MCSS2001) and Related Topics,* Oberpfaffenhofen, Germany, September 26–28, 2001, pp. 1–4.

[25] Hane, S., et al., "OFDM Null Steering Array Antenna (in Japanese)," *Technical Report of IEICE,* RCS2002-124, July 2002, pp. 31–36.

[26] Budsabathon, M., et al., "On Pre-FFT OFDM Adaptive Antenna Array for Delayed Signal Suppression," *Technical Report of IEICE,* RCS2002-125, July 2002, pp. 37–42.

[27] Raleigh, G. G., and J. M. Cioffi, "Spatio-Temporal Coding for Wireless Communication, "*IEEE Trans. Commun.,* Vol. 46, No. 3, March 1998, pp. 357–366.

[28]   Tarokh, V., N. Seshadri, and A. R. Calderbank, "Space-Time Codes for High Data Rate Wireless Communication: Performance Criterion and Code Construction," *IEEE Trans. Inform. Theory,* Vol. 44, No. 2, March 1998, pp. 744–765.

[29]   Tarokh, V., H. Jafarkhani, and A. R. Calderbank, "Space-Time Codes from Orthogonal Designs," *IEEE Trans. Commun.,* Vol. 45, July 1999, No. 5, pp. 1456–1467.

[30]   Li, Y., N. Seshadri, and S. Ariyavisitakul, "Channel Estimation for OFDM Systems with Transmit Diversity in Mobile Wireless Channels," *IEEE J. Select. Areas Commun.,* Vol. 17, No. 3, March 1999, pp. 461–471.

[31]   Sampath, H., et al., "A Fourth-Generation MIMO-OFDM Broadband Wireless System: Design, Performance, and Field Trial Results," *IEEE Commun. Mag.,* Vol. 40, No. 9, September 2002, pp. 14–149.

[32]   Salz, J., "Digital Transmission Over Cross-Coupled Linear Channels," *AT&T Technical Journal,* Vol. 64, No. 6, July–August 1985, pp. 114–1,159.

[33]   Andersen, J. B., "Array Gain and Capacity for Known Random Channels with Multiple Element Arrays at Both Ends, " *IEEE J. Select. Areas Commun.,* Vol. 18, No. 3, Nov. 2000, pp. 2172–2178.

[34]   Ochiai, H., and H. Imai, "Performance of the Deliberate Clipping with Adaptive Symbol Selection for Strictly Band-Limited OFDM Systems," *IEEE J. Select. Areas Commun.,* Vol. 18, No. 3, November 2000, pp. 2270–2277.

[35]   Wight, J., "RF Architectures for OFDM," *IEEE RAWCON 2002,* Sunday Workshop Notes on Next Generation Wireless LAN Systems, Boston, MA, August 11–14, 2002.

[36]   Cox, D. C., "Linear Amplification with Nonlinear Components," *IEEE Trans. Commun.,* Vol. 23, No. 12, December 1974, pp. 1942–1945.

[37]   Chiereix, H., "High Power Outphasing Modulation," *Proc. IRE,* Vol. 23, No. 11, November 1935, 1370–1392.

[38]   Nagareda, R., K. Fukawa, and H. Suzuki, "A Calibration Method of Amplitude and Phase Balance of Power Amplifiers for LINC by Least-Squares Method (in Japanese)," *IEICE Technical Report,* RCS2002-169, November 2001, pp. 7–12.

[39]   Gu, J., "An Enhanced Linear Amplification with Non-Linear Component (ELINC) System," *Proc. IEEE RAWCON 2002,* Boston, MA, August 11–14, 2002, pp. 173–176.

List of Acronyms

1G	first generation
2G	second generation
3G	third generation
3GPP	Third Generation Partnership Project
4G	fourth generation
ACI	adjacent channel interference
A/D	analog to digital
ADSL	asymmetric digital subscriber lines
AM	amplitude modulation
ARIBE	Association of Radio Industries and Businesses
AWGN	additive white Gaussian noise
BDMA	band division multiple access
BER	bit error rate
BPSK	binary phase shift keying
BRAN	broadband radio access network
BWA	broadband wireless access
CDM	code division multiplexing
CDMA	code division multiple access
CPSK	phase shift keying/coherent detection
CSMA/CA	carrier sense multiple access with collision avoidance
CW	continuous wave
D/A	digital to analog
DAB	digital audio broadcasting
DFS	dynamic frequency selection

DPSK	phase shift keying/differential detection
DS	direct sequence
DSL	digital subscriber line
DVB-T	terrestrial digital video broadcasting
DFT	discrete Fourier transform
E_b/N_0	ratio of energy per bit to power spectral density of noise
EGC	equal gain combining
ETSI	European Telecommunications Standards Institute
FDP	frequency domain pilot
FEC	forward error correction
FFT	fast Fourier transform
FH	frequency hopping
FM	frequency modulation
GSM	global systems for mobile telecommunications
GSN	generalized shot noise
HDR	high data rate
HF	high frequency
Hi-Fi	high fidelity
HIPERLAN/2	High Performance Radio Local Area Network Type Two
HSDPA	high-speed downlink packet access
IDFT	inverse discrete Fourier transform
IFFT	inverse fast Fourier transform
IEEE	The Institute of Electrical and Electronic Engineers, Inc.
i.i.d.	independent and identically distributed
IMT	international mobile telecommunications
IS	interim standard
ISDB-T	terrestrial integrated services digital broadcasting
ISI	intersymbol interference
ISM	industrial, scientific and medical
ITU	International Telecommunication Union
ITU-R	International Telecommunication Union-Radiocommunication Standardization Sector
LAN	local area network
LINC	linear amplification with nonlinear components
LMS	least mean square
MAC	medium access control
MAN	metropolitan area network

MC-CDM	multicarrier code division multiplexing
MC-CDMA	multicarrier code division multiple access
MCM	multicarrier modulation
MFN	multifrequency network
m.g.f.	moment generating function
MIMO	multiple input and multiple output
MMAC	multimedia mobile access communication
MMSEC	minimum mean square error combining
MRC	maximum ratio combining
MUI	multiple user interference
OFDM	orthogonal frequency division multiplexing
OFCDM	orthogonal frequency and code division multiplexing
OFDMA	orthogonal frequency division multiple access
ORC	orthogonality restoring combining
PAN	private area network
PAPR	peak to average power ratio
PDC	personal digital cellular
p.d.f.	probability density function
PDNR	preliminary draft of new recommendation
PHY	physical layer
PM	phase modulation
PN	pseudo noise
QAM	quadrature amplitude modulation
QPSK	quadrature phase shift keying
RMS	root mean square
SCM	singlecarrier modulation
SF	spreading factor
SIC	serial interference cancellation
SNIR	signal to noise plus interference power ratio
SNR	signal to noise (power) ratio
SS	spread spectrum
SSPA	solid state high power amplifier
TDMA	time division multiple access
TDMA/DSA	time division multiple access with dynamic slot assignment
TDP	time domain pilot
TPC	transmission power control
US	uncorrelated scattering
VSF	variable spreading factor

| WSS | wide sense stationary |
| WSSUS | wide sense stationary uncorrelated scattering |

About the Authors

Shinsuke Hara was born in Osaka, Japan, on January 22, 1962. He received his B.A., M.A., and Ph.D. degrees from Osaka University in Osaka, Japan, in 1985, 1987, and 1990, respectively.

From April 1990 to September 1997, he was an assistant professor at the Department of Communications Engineering, Faculty of Engineering, at Osaka University. Since October 1997, he has been an associate professor with the Department of Electronic, Information System, and Energy Engineering at the Graduate School of Engineering at Osaka University. From 1995–1996, he was a visiting scientist with the Telecommunications and Traffic-Control Systems Group at Delft University of Technology in Delft, the Netherlands.

Dr. Hara's research interests include the application of digital signal processing techniques for high-speed and high-reliable wireless communications systems. He was the technical program cochairman of the PIMRC'99 International Symposium in Osaka, Japan, as well as the technical program cochairman of the WPMC'01 International Symposium in Aalborg, Denmark.

Ramjee Prasad was born in Babhnaur (Gaya), India, on July 1, 1946. He is now a Dutch citizen. He received his B.A. in engineering from the Bihar Institute of Technology in Sindri, India, in 1968, and his M.A. in engineering and Ph.D. from Birla Institute of Technology (BIT) in Ranchi, India, in 1970 and 1979, respectively.

He joined BIT as a senior research fellow in 1970 and became an associate professor in 1980. While at BIT, Dr. Prasad supervised a number of research projects in the area of microwave and plasma engineering. From 1983–1988, he was with the University of Dar es Salaam (UDSM), in Tanzania, where he became a professor of telecommunications in the Department of Electrical Engineering in 1986. At UDSM, Dr. Prasad was responsible for the collaborative project Satellite Communications for Rural Zones with Eindhoven University of Technology in the Netherlands. From February 1988–May 1999, he was with the Telecommunications and Traffic Control Systems Group at Delft University of Technology (DUT), where he was actively involved in the area of wireless personal and multimedia communications (WPMC). He was the founding head and program director of the Center for Wireless and Personal Communications (CEWPC) of International Research Center for Telecommunications–Transmission and Radar (IRCTR).

Since June 1999, Dr. Prasad has been with Aalborg University as the research director of the Department of Communication Technology and holds the chair of wireless information and multimedia communications. He was involved in the European ACTS project Future Radio Wideband Multiple Access Systems (FRAMES) as a DUT project leader. He is currently the project leader of several international, industrially funded projects.

Dr. Prasad has published more than 300 technical papers, contributed to several books, and has authored, coauthored, and edited 12 books. These books, all published by Artech House, are: *CDMA for Wireless Personal Communications*; *Universal Wireless Personal Communications*; *Wideband CDMA for Third Generation Mobile Communications*; *OFDM for Wireless Multimedia Communications*; *Third Generation Mobile Communication Systems*; *WCDMA: Towards IP Mobility and Mobile Internet*; *Towards a Global 3G System: Advanced Mobile Communications in Europe, Volumes 1 & 2*; *IP/ATM Mobile Satellite Networks*; *Simulation and Software Radio for Mobile Communications*; *Wireless IP and Building the Mobile Internet*; and *WLANs and WPANs towards 4G Wireless*. His current research interests lie in wireless networks, packet communications, multiple-access protocols, advanced radio techniques, and multimedia communications.

Dr. Prasad has served as a member of the advisory and program committees for several Institute of Electrical and Electronic Engineers, Inc. (IEEE) international conferences. He has also presented keynote speeches and delivered papers and tutorials on WPMC at various universities, technical institutions, and IEEE conferences. In addition, he was a member of the European cooperation in the scientific and technical research (COST-231) project dealing with the evolution of land mobile radio (including personal) communications as an expert for the Netherlands, and he was a member of the COST-259 project. He was the founder and chairman of the IEEE Vehicular Technology/Communications Society Joint Chapter, Benelux Section, and is now the honorary chairman. In addition, Dr. Prasad is the founder of the IEEE Symposium on Communications and Vehicular Technology (SCVT) in the Benelux. He was the symposium chairman of SCVT'93.

Dr. Prasad is the coordinating editor and editor-in-chief of the *Kluwer International Journal on Wireless Personal Communications* and a member of the editorial board of other international journals, including the *IEEE Communications Magazine* and the *IEE Electronics Communication Engineering Journal.* He was the technical program chairman of the PIMRC'94 International Symposium, held in The Hague, the Netherlands, from September 19–23, 1994, and also of the Third Communication Theory Mini-Conference in Conjunction with GLOBECOM'94, held in San Francisco from November 27–30, 1994. He was the conference chairman of the 50th IEEE Vehicular Technology Conference and the steering committee chairman of the second International Symposium WPMC, both held in Amsterdam, the Netherlands, from September 19–23, 1999. He was the general chairman of WPMC'01, held in Aalborg, Denmark, from September 9–12, 2001.

Dr. Prasad is the founding chairman of the European Center of Excellence in Telecommunications, known as HERMES. He is a fellow of IEE, a fellow of IETE, a senior member of IEEE, a member of the Netherlands Electronics and Radio Society (NERG), and a member of IDA (Engineering Society in Denmark).

Index

For further information on these and other Artech House titles, including previously considered out-of-print books now available through our In-Print-Forever® (IPF®) program, contact:

Artech House
685 Canton Street
Norwood, MA 02062
Phone: 781-769-9750
Fax: 781-769-6334
e-mail: artech@artechhouse.com

Artech House
46 Gillingham Street
London SW1V 1AH UK
Phone: +44 (0)20 7596-8750
Fax: +44 (0)20 7630-0166
e-mail: artech-uk@artechhouse.com

Find us on the World Wide Web at:
www.artechhouse.com